Fail Fight Finish

Gregory Archbold

Published by **Ryker Blueprint Press**

This book is intended for informational and educational purposes only. The author and publisher make no representations or warranties with respect to the accuracy or completeness of the contents and specifically disclaim any implied warranties of fitness for a particular purpose. The advice and strategies contained herein may not be suitable for every individual or situation. Readers are responsible for their own actions and decisions.

ISBN: 978-1-969278-06-8

Library of Congress Control Number: [Optional / Insert if applicable]

Printed in the United States of America

RYKER
BLUEPRINT
PRESS

Contents

Acknowledgements 1

Introduction 3

1. Before the Uniform 6

2. What I Do Not Change, I Choose 14

3. The Long Road to Commissioning 20

4. Why Real Leadership Has Nothing to Do with Position 24

5. The Leadership Trifecta 40

6. The Power of Silence 50

7. Coach, Mentor, Leader 57

8. From Peer to Leader 66

9. Leadership Discretion 74

10. Excellence by Design 86

11. Legacy or Limitation? 92

12. The Moment Leadership Is Tested 98

13. Creating Change in an Organization 108

14. Creating Value in Your Organization 116

15. The Empty Slice 124

16. Preparation Over Panic 133

17. The Man in the Mirror 141

18. What Gets Checked Gets Done   149

19. React Less, Lead More   157

Conclusion   178

Glossary of Abbreviations   181

Acknowledgements

No book is written alone, and this one is no exception.

First and foremost, I owe everything to my wife, Kathy. For years, she saw this book long before it ever existed on paper. She listened to the stories more times than she probably cared to, encouraging reflection, and pushed me when I hesitated. Kathy is not known for sugarcoating; whether encouragement or criticism, she challenged me to write when I doubted whether the words mattered. She carried the weight of countless late nights, extended absences, deployments, transitions, and uncertainty, all while believing in me more than I believed in myself. This book exists because of her patience, her strength, and her unwavering support. Kathy, thank you for walking beside me through every season and for reminding me that the stories were worth telling.

To my children, Amy, Myles, Javier, and Masaya, you have been my constant reminder of why leadership matters beyond rank, title, or success. You pushed me not by words, but by knowing you were watching made me want to be the best father I could be. Everything I teach about leadership begins at home. My hope has always been that you would never have to look far for an example of integrity, accountability, and effort. If this book carries any truth, it is because I tried to live it first for you.

I am also profoundly grateful to my sister-in-law, Angela, who volunteered her time, energy, and support throughout this journey. She stepped in whenever help was needed, often without being asked, and never sought recognition for it. Her willingness to assist behind the scenes made this process lighter and more manageable than it would have been otherwise. Angela, thank you for always showing up.

I am equally grateful to the many leaders who shaped me along the way: Cecil Archbold, Maria Archbold, Geraldine Ronan, William Franklin, Kevin Tillman, Thomas Perez, Stan McChrystal, Tony Thomas, Michael Gragg, Jim Phillips, Robert Lutz, Brett Ackermann, Darcy Overbey, Chris Hodl, and Michael Brimage, to name just a few. Some coached me when my skills were lacking. Some of these leaders mentored me when my perspective was narrow. Others led me when the pressure was high, and the consequences were real. Each one, whether they knew it at the time or not, left a mark. You showed me what right looks like, what courage sounds like, and what responsibility feels like. Any leadership wisdom found in these pages is a reflection of lessons first demonstrated by you.

To the soldiers, teammates, colleagues, and professionals I had the privilege to serve alongside: thank you for trusting me, challenging me, and holding me accountable. Leadership is a shared experience, and I learned as much from those I led as from those who led me. Finally, to the reader: if any part of this book challenges you, encourages you, or causes you to pause and reflect, then the effort was worth it. Leadership is a relay, not a race. I hope that what was given to me through others now helps you carry the standard forward for yourself and for those who will one day follow you.

Introduction

Growing up, I always knew I was a little different. I am not sure if it was the aroma of fried plantains that wafted through my childhood home, while just down the street, the vinegar tang of ketchup accented the burgers and fries that my friends ate. Or maybe it was church. My friends told stories about choirs that sang for hours on Sundays, stretching well into the afternoon. For me, the Catholic Mass was over and done within the hour, no drawn-out sermons, no long list of hymns. When my friends talked about the long church hours, I often just nodded in agreement, quietly making up my own church stories so that I would not stand out too much from the rest of the crowd.

But the differences went deeper than food or church. I had both parents living at home; most of my friends had single mothers. I knew my father and had a good relationship with him; some of my friends had never met theirs. For a family tree project in school, I traced my roots back to the Latin American country of Panama. Most of my classmates drew branches that reached down into the American South. One boy, who knew almost nothing about his father, was told by his mother to "just make something up."

That day, I had the unusual experience of feeling extremely proud of my heritage, while at the same time being painfully aware that my home life somewhat excluded me from my circle of friends. It made me uncomfortable. Being a kid was hard enough. Being a different kid? Well, there were times when simply getting through the day as an outsider was an absolute struggle. Sadly, I discovered that a stable, two-parent home could sometimes translate into a lonely existence. Identity can be complicated when the outside world around you does not match the life inside your home. I straddled two cultures, two sets of expectations, and two ways of life. At times, I was not at all sure where I truly belonged.

To cope with the uncertainty, I told little lies to blend in, hiding the pieces of myself that made me stand out like a sore thumb. But over time, those same differences gave me a gift. They trained me to notice people, to pay attention to the small details that others overlooked. This power of observation allowed me to step into rooms where I did not look or sound like everyone else and still find a way to belong. The truth of the matter is that those differences and early struggles laid the foundation for my adult life. They were not obstacles that held me back; they were building blocks that taught me resilience and adaptability. They instilled in me the courage to forge my own path and become a leader. Leadership is not born from ease or convenience; it arises from the ability to navigate discomfort, to stand in the tension of being different, and to find strength in that tension.

I think you must know this part of my story. Without those uncomfortable beginnings, without the challenges that forced me to grow even when I was sure I was not ready, I would never have become the leader that I am today. I believe it is my duty to share this vital discovery with you. This book is about leadership, yes, but not in the abstract way leadership is often discussed. It is not a list of theories or principles plucked from some sterile, generic manual. It is about the real, messy, and utterly human journey of becoming a leader. It is shaped by family, culture, adversity, failure, and every single victory, no matter how small, incurred along the way. These are the lessons I have learned through trial and error, the experiences that left their marks on me, and the insights gained from it all. I want to share my stories of leadership in the making with others who carry the weight of a team, a unit, a military outfit, or a civilian organization. Even if the weight you are carrying is just your own, and you want to do it better, this book is for you.

I hope that as you turn these pages, you will see your own story reflected, that you will come to see the struggles you face not as barriers but as raw material necessary for your growth. That the weight of being different or the simple awkwardness of feeling out of place will start to look less like a burden and more like the very thing that equips you to lead. Because that is the paradox of leadership: the challenges that once made you feel small often become the same sources of strength that inspire others.

This book is my personal journey, but it is also an invitation for you to embark on a trip of your own. You have everything you need; it is just a matter of knowing where to look.

Chapter One

Before the Uniform

Leadership at My Kitchen Table

B efore I understood leadership, before I learned to serve or inspire others, I was being shaped quietly, steadily, and without fanfare by two people whose names will never appear in history books but whose influence echoes in every decision I make. My parents did not raise me with grand speeches or elaborate philosophies. They did not sit me down and lay out a master class in leadership. But they taught me, through example, through sacrifice, with quiet strength, unwavering discipline, and a love that ran deeper than any words could hope to express. They did not always give me what I wanted, but they always gave me what I needed. They modeled a principle I would later come to live by as a leader: give your people truth, not comfort; give them a purpose, not just provision; and above all, give them your presence.

My mother's journey began in hardship: born into poverty, raised in the shadows of broken systems, and pushed into survival mode at far too young an age. From that place of lack, she somehow created abundance for her children. She taught me what it meant to fight for something better, not just for yourself, but for those who come after you. She showed me that resilience is not loud or forceful; it is steady and persistent.

My father's story is equally powerful. Rooted in community and duty, he became the man of the house when his father passed away. There was no ceremony to mark the shift, no applause, no promotion, just the silent assumption of responsibility, carried out with integrity, gratitude, and the hope that his children might one day go further than he ever could.

He was rooted in community and duty. In those days, community meant something more profound than mere proximity. It was neighbors who knew each other's names, who showed up when a roof needed mending or a meal needed to be shared. My father understood that strength was not about standing alone; it was about standing *with* others. He learned early that service to your community was not optional; it was expected.

Duty, for him, began at home. He rose before the sun to help his mother with chores, went to school, worked odd jobs after class, and still made sure his younger siblings were fed, dressed, and on the right path. There was no complaining, no talk of sacrifice, just a quiet resolve that this was what a man did when his family needed him.

What I admired most was how he carried that same sense of duty into every corner of his life. He treated his word like a promise. He showed up when others did not. He looked out for people who could not repay him. It was not about being noticed; it was about doing right by others because that is what his father had done before him.

Looking back, I realize that his leadership began long before he ever used the word "leader." It was born in those small acts of service, in the way he shouldered responsibility without fanfare. His example taught me that authentic leadership does not need recognition; it needs reliability.

And that is the legacy he passed on: community gives us purpose, and duty provides that purpose with direction.

My dad's journey was not his alone. He had three brothers, and together they made one of the hardest choices of their lives: leaving their homeland, a place rich with family familiarity and identity, and starting over in a country that promised opportunity but guaranteed nothing.

My grandmother led the way, leaving first to build a foundation. My father followed soon after, while the other three brothers stayed behind until he was able to bring them over.

It was a plan rooted in faith and sacrifice, one generation paving the way for the next. When they arrived in the United States, they discovered that success here came with its own set of challenges. The barriers they faced had little to do with ability and everything to do with perception. They encountered racism, stereotyping, and the sting of being underestimated because of an accent or last name. Intelligence and work ethic were often questioned before they had a chance to prove themselves. Still, they endured. They refused to let someone else's limited view define their potential. Every insult, every closed door, every quiet act of exclusion became fuel, a reminder of why they had come in the first place. It was a heavy cost to pay for a better future, but they paid it willingly, believing that their sacrifices would open doors for their children that had never been opened for them. Through persistence, faith, and the unbreakable bond of family, they built a life defined not by circumstances but by courage.

This chapter is a tribute to them, not just as my parents but as individuals with their own dreams, battles, losses, and triumphs. Their lives did not just influence mine; they built it. To understand who I became as a leader, I must begin by knowing who they were.

Mom's Story: Strength in Scarcity

Maria grew up with very little. Many days, one meal was a luxury; some days, there was no meal at all. My maternal grandfather battled alcoholism, leaving the family constantly struggling to survive. To keep everyone afloat, my grandmother, whom we lovingly called Connie, worked as a domestic cleaner for Americans living in the Panama Canal Zone. There was no space for extravagance: no vacations, no outings, just subsistence, structure, and love.

When my grandparents divorced, the three siblings were split among three different extended family homes. It was painful, but not uncommon. My grandmother wanted more for her children, so she took a leap of faith and left Panama for Chicago to pursue education and opportunity. She worked tirelessly, sending money when she could. From this, my mother learned one of life's harshest truths: success is not given; it is earned through persistence, sacrifice, and sheer will.

My mom met my father one afternoon while shopping at the commissary, the grocery store located on the Pacific side of Panama, where only those living in the Canal Zone were permitted to shop. She was with a friend who happened to know one of my dad's friends and thought the two of them should meet. That casual introduction would change the course of both of their lives. After years of courtship, they followed my grandmother to Chicago in search of a better life. There was no road map; just grit and determination in those early years. Mom faced a steep learning curve as she adjusted to a new culture and a new way of life. Still, she refused to let that stop her. She pursued her education not only as a personal goal but as a declaration that our family would not remain trapped in the same cycle of poverty and limitations that had defined her upbringing. She set a new standard for what was possible, and doing so, she did not just change her own life; she raised the bar for all of us who would follow. She would not allow her family to remain stuck in the cycle of poverty and limitation that she had grown up in; she set a new standard, not only for herself but for all of us.

When she became a mother, everything became intentional. Homework was not optional. Nutritious meals were non-negotiable. Discipline was consistent. Education, structure, and personal responsibility were not rules; they were family values. All three of her sons would eventually earn master's degrees, a testament to the importance of education in the Archbold family.

Some stories are difficult to tell, not because they are tragic, but because the people in them never asked to be seen as the story's main characters. That is how my mom was. She did not jump up and down on a trampoline, saying, "Look at me! What a great mom I am!" She simply did the work without fanfare or accolades. In this way, she held a family together and set a high bar without saying a word.

My mom was a respiratory therapist by trade at Northwestern Memorial Hospital in downtown Chicago. Our home was in Oak Park, one of Chicago's many suburbs. Even though we were a two-car family, parking downtown was expensive, so my mom used the Chicago Transit Authority (CTA) to get to and from work each day. Her commute went something like this: (1) leave the house and walk a block to the bus stop on the corner of Austin Blvd. and North

Avenue; (2) take the bus down to the Lake Street station; and (3) catch the elevated train, or "El" for short.

The El was not just transportation; it was part of daily life for many people who lived in and around the city. You would hear it before you saw it, the rumble of its progress echoing between buildings, the screech of metal on metal as it made stops. It ran overhead, pulsing like a steel heartbeat as it traversed the city, connecting neighborhoods and lives, one rattling car at a time. The length of Mom's trip depended on the time of day and whether the CTA was on schedule. It took about an hour and fifteen minutes on a good day, culminating in a fifteen-minute walk to the hospital.

During the summer, it was tolerable; however, it tested anyone's resolve in the cold of a Chicago winter. I still remember watching her get ready as she put on her long coat, scarf, and boots, always with a smile. My mom repeated this daily and took pride in never being late. She never complained, never mentioned how cold the walk was in January or how draining it must have been to stand during rush hour after a twelve-hour shift. She just did it. Mom's example showed my brothers and me the importance of working through situations that made us uncomfortable. It was a leadership lesson I did not even know I was learning until many years later.

She always said, "When my sons were born, I poured everything into them." It was more than a statement; it was a way of life.

Her greatest source of pride was seeing her children thrive. For me specifically, it was when I joined the United States Army. I believe it is because she knew that in the Army, the discipline she had sown into me would finally take root and bloom into fruition.

"If I am remembered for anything," she once told me, "let it be for loving unconditionally."

She dreams of gathering all of her grandchildren and taking them on a cruise, surrounded by the laughter, resilience, and love that define our family legacy. If that moment ever comes, I will know it is not just a family vacation; it is a full-circle moment for a woman who built a future from nothing.

Lessons from My Mother

- Fight for the future, even when your present is bleak.

- Give people consistency, and they will grow in it.

- Lead through love, not volume.

Dad's Story: Duty Without Applause

For as long as I can remember, my dad was a living, breathing lesson in time management and hard work. In the early 1990s, a popular comedy show called *In Living Color* featured one of its funniest recurring skits titled, "*Hey Mon.*" It poked fun at West Indian families who always had multiple jobs: "I am a doctor, a lawyer, and a cab driver!" they would say in heavy Caribbean accents, stacking roles with pride and exhaustion. The sketch was satire, but in our house, it was just Tuesday.

That was my father. Always working, doing as many jobs as he needed to provide for his family. At one point, my dad was working for Cory Food Services, pursuing his MBA at Roosevelt University, working at Debbie Howell Cosmetics, managing the apartment building we lived in (plus two others), and teaching night classes at a community college. He did not just work 9:00 a.m. to 5:00 p.m.; he worked 6:00 a.m. to 9:00 p.m., six days a week, sometimes seven.

It was not uncommon to see him come home after a full day at Cory, pull out the paperwork for Debbie Howell, then shift gears again to go through the rent ledgers, prepping the deposit slips for the buildings. It was done at the table, with no grumbling or drama. It was another in a long line of nights spent wearing three hats and juggling chainsaws without blinking.

It was never a big production. Dad did not live for applause. He got up and did what needed to be done every day, without exception. He never really talked to us about money, but I imagine that, in the early days, his hustle was at least partly about keeping the lights on and paying the rent. Still, as I got older, I realized it was about more than finances. It was about helping people, making a difference, and living a useful life.

One Saturday comes to mind. I was over at a friend's house, and both his parents were at home relaxing. I looked at the difference between Saturday at my friend's house and Saturday at mine. I could not help but feel a little jealous of my friend and wonder, *why is my dad always gone?* I did not understand it then, but I do now.

Please do not misunderstand my meaning here. There is nothing wrong with spending Saturday at home relaxing. That is not at all what I am saying. What I am saying is that my father was not at home relaxing, because he was out building something. Not just a career but a legacy. Not solely for his family but for everyone he worked with and for those he taught. This is not achieved by the simple marking of time, clocking in and out, and putting in forty hours per week. Success like that which my father sought is achieved not by being average but by going above and beyond. It takes leadership. My father was leading by example, by endurance, and by purpose.

The consistent work ethic, commitment to improvement, and servant-hearted leadership my father showed me as a child shaped how I lead today. I used to think leadership meant standing in front of people and giving direction through orders. But my dad taught me it is also about showing up even when no one is watching, being steady and sound, and finishing what you start. Lead with humble consistency. Be useful more than impressive. And remember: impact is not built in a day; it is built daily.

My father's values were clear: live within your means, be kind, love deeply, laugh often. And always leave a place better than you found it. If he could be remembered for anything, it would be for raising his family with intentional love and responsibility.

Lessons from My Father

- Do not wait for the title you want; step into the role.

- Provide stability without seeking reward.

- Teach what you know, give what you have, and never stop praying for those you lead.

Final Reflection

I have led teams into combat, given briefings to generals, and spoken on stage many times. But nothing I have done as a leader means more than this truth: I stand on the shoulders of two people who never made the news but have made a lasting difference, not only in my life and my two brothers' lives but also in the lives of *many* others. Their leadership was not about being in charge; it was about being present, choosing love over comfort, discipline over ease, and responsibility over recognition.

So when people ask where my leadership journey began, I do not point to a military school, a badge, a job title, or a social media influencer. I point to my parents' kitchen table, where my mom created masterpieces like hot dogs, pork and beans, and white rice, still a favorite dish. It is the same table where my dad gave us lectures over family meals, preparing us for life. Most importantly, I point to their quiet sacrifices and the foundation they laid, stone by stone, lesson by lesson. All day, every day.

Mom and Dad, if you are reading this, I cannot tell you how much you are appreciated and loved. You did not just raise me; you built me.

Chapter Two

What I Do Not Change, I Choose

I was not just a typical restless teenager who thought he could do as he pleased; I was more of an explosive storm, like the Tasmanian Devil who chased Bugs Bunny in those old cartoons. I was looking for something to destroy, and most days that something was me. Restless and lost, I turned to heavy drinking and drugs, surrounding myself with the kind of people who encouraged my worst impulses. I told myself I was having fun, but deep down, I was running from expectations, responsibility, and myself.

By the grace of God, I managed to graduate from high school. However, I had no real plans for my future, and I knew my parents were aware of this. They loved me, but they would not enable my reckless behavior. When I finally pushed them too far, I did what I had always done: I ran.

At first, it felt like freedom, no rules, no lectures about my future, just an endless party. I crashed at friends' houses, convinced I could keep up with this lifestyle forever. But reality has a way of catching up. One by one, my friends' parents started asking the dreaded question no couch surfer wants to hear: "When are you going home?" Eventually, I had nowhere to go. I slept in a friend's garage, on the streets, or in the woods behind an apartment complex where I had an acquaintance.

One night still sticks with me: I was partying with some friends. One friend's parents had gone away for the weekend, and we were left to ourselves without any adult supervision. I also knew I would have a good bed for a few nights, and that eased my mind. In short, I knew the next couple of days were going to be epic. In typical teenage fashion, we called a few people, which turned into a few more people, and the party was on. You do not need a vivid description; you were once a teenager yourself. You can imagine the debauchery that went on. But Sunday came fast. We needed to clean up and get the house back to normal. More importantly, I had to leave. I played it off and told my friends, many of whom were unaware of my current living situation, that I was "going home," and I left.

I found myself on the streets again. The few friends who were always a good backup were unavailable. With no other options, I headed back to the woods behind the apartments. I knew I would be safe there if I left before the morning dog walkers arrived. It was early fall, and although the leaves had not yet entirely changed and fallen, it was already cold. The wind was blowing, but I was able to use some bushes to shield myself from its force. Thank God I had a hoodie. I could hear every acorn fall, every branch sway in the wind. Even though I was not afraid of wild animals, my mind still played unsettling tricks on me. Every noise had me wondering, *What is that?* I did not get much sleep, and the night seemed to drag on and on. I remember finally seeing the sunrise, then getting up and walking to yet another one of my friends' houses, where the parents would soon be leaving for work. After they did, I would have an opportunity to shower, eat, and steel myself before going through the cycle again.

I told myself I was fine, that I had everything under control. But the truth? I had no direction, no future. I kept sliding down the wrong side of life, the one leading me straight toward destruction. Something had to break, and it did. It was not a dramatic event, no family intervention, no overdose, just a moment of raw clarity. I was sitting alone, the high wearing off, my body aching, my mind racing. For the first time, I let myself admit the truth: I could not keep living like this.

With hindsight being 20/20 and through developing a deep, personal relationship with Jesus, I now know that this moment of clarity was His way of telling me to get my affairs in order.

Checking myself into treatment was the hardest decision I had ever made. Rehab was not an easy fix; it was brutal. It stripped away all my distractions and excuses, forcing me to look at the person I had become. Group therapy introduced me to people whose stories mirrored my own, and I realized I was not as alone as I thought. I also had the opportunity to see what my life would become if I kept on the same path. If you are open and receptive to the truth of recovery, there will always be one moment that hits you like a gut punch, and for me, it was not my own story. It was AJ's.

We were in a group session when he told us how he got drunk one night, drove off in a haze, woke up in a jail cell, was charged with vehicular homicide, and was sentenced to fifteen years in the penitentiary for one wrong decision. One moment of being numb to the world and now he had to live with the fact that someone else did not wake up because of him and that he would have given up fifteen years of his life.

I did not say a word. I just sat there, frozen, because the truth was that it could have been *me*. I had driven high. I had driven drunk. I had blacked out and laughed it off like it was nothing. But hearing AJ say those words and seeing the weight he carried, hit differently. I did not feel judged. I felt fear. Gratitude. And a cold wave of realization that if I did not change, I was going to end up in a place I could not climb back from. That was my turning point.

It was not a counselor. It was not a workbook. It was AJ's pain. His honesty. His consequence. That is when I stopped pretending that I was going through a phase or that I had time to figure things out. I did not.

That was when I told myself, "Get your stuff together, or you will break something you cannot fix." That realization caused something in me to crack, but not in a way that made me weaker. It was like the shell surrounding two sides of me finally broke, and I could at last clearly see the man I did not want to become and the one I still had a shot at becoming.

That was the day I chose to change.

I had a counselor named Jill who told me to confront the pain, and I did. She told me to be honest about my feelings, and I did. The environment at Gaudenzia gave me discipline, something that, until then, I had never truly embraced.

However, the most significant transformation did not come from therapy or structure but from within. Rehab was not just about alcohol and drug abuse. It taught me how to think differently, and it rewired my perspective on myself and the world around me. Together with this information and the guidance my parents provided, I put myself on a different trajectory. With the counselors' help, I realized for the first time that my destructive actions had consequences. I could not blame my friends or my parents for my circumstances; I could not blame my desire to fit in for my situation. All I could do was look in the mirror. I now recognize that *I* have the power to change my life.

I remember the moment it truly clicked: *What I do not change, I choose.*

It was up to me to either rewrite my own story or allow myself to become another statistic of a young Black man falling into the cracks of despair. I knew my thinking had to change, and as it did, it became my compass, marking the beginning of a new path to self-leadership, one grounded not in authority, but in accountability, humility, and the courage to create my own narrative. It became the foundation of my leadership journey, the turning point that would shape my future.

Recovery taught me that **leadership does not start with a title; it begins with how you lead yourself when no one is watching**. Before I could ever lead others, I had to learn how to lead myself.

In those early days, self-leadership did not look the least bit glamorous. It looked like waking up on time when I did not feel like it. It looked like going to bed at a reasonable hour. It looked like leaving the house early so I could make the walk to my job and never be late. It looked like showing up for the group when I wanted to hide, saying "no" when my old habits whispered "yes." It looked like running into old "friends" and telling them I was not available so I would not fall back into those destructive habits that nearly broke me. It looked like paying my share of the rent on time and with no excuses, just like my roommates did. All of this was necessary because if I could not manage the basics, if I could not build a foundation, I had no business dreaming about anything bigger.

I started building a routine, not because it felt good, but because I needed structure to survive. I was no longer living under my parents' roof, so I had to make my own choices and live by them. Every morning, I made a point of naming three simple things:

1. What I was grateful for

2. One thing I wanted to do better than the day before

3. One person I wanted to encourage, even if it was just by saying hello

These thoughts became my touchstone, a way of taking responsibility for how I present myself in life.

But let me be honest with you, it was not flashy. There were no grand speeches. No applause. No motivational soundtrack playing in the background. Most days, self-leadership was just doing the next right thing, over and over, until the hard stuff started to feel normal. I was not chasing motivation. I was building muscle memory. I had to learn how to make good decisions when nobody was around to see them and when there was no instant reward waiting on the other side.

That is the part nobody talks about. Change is quiet. Progress is boring. But if you stay with it long enough, boring becomes powerful. Some days, just getting out of bed was a win. On other days, I would catch myself advising a new person in the group and realize I was becoming someone people looked to. Not because I had all the answers, but because I was consistent. I showed up. I kept my word. And in a world full of empty promises and flaking out, that started to matter more than I ever expected.

It was not glitzy, but it was real. Authentic leadership that lasts begins with the small, unseen choices. In the places where discipline feels hard and growth feels slow. That is where self-leadership is built. That is where I rebuilt myself.

Over time, self-control evolved into quiet confidence. I was not perfect, but I was intentional. I learned to delay gratification, to be comfortable with discomfort, and to take ownership of my choices without making excuses or blaming others.

It was a turning point in my life. Once I learned how to lead myself with discipline, compassion, and honesty, I could finally show others how to do the same. Whether it was a friend in recovery, a teammate, or, eventually, soldiers under my command, they did not just follow my words; they followed my example.

Self-leadership taught me how to earn trust rather than demand it. It taught me how to listen actively, not just react reflexively. And most of all, it taught me that **authentic leadership always starts from within**.

The most dangerous place we can be is where we feel comfortable self-destructing. I know; I lived there. It is easy to believe we are too far gone, that our past disqualifies us from a better future and the ability to be great. But that is a lie the devil tells us to keep us mired in muck.

The truth is that rock bottom is not a place; it is a decision. It is when you stop pretending everything is fine and admit you want something more. That is where transformation begins, not in the absence of pain but in the presence of truth.

I used to think leadership was about *telling* people what to do. Now I know it is about *showing* people what is possible by how you live, rise, and own your past without being defined by it.

And so, I leave you with this: If you are waiting for permission to change, this is it. If you are waiting for a sign, this is it. What you do not change, you choose. And what you choose becomes your legacy. Make it one worth remembering.

Chapter Three

The Long Road to Commissioning

A Lesson in Loopholes, Grit, and Leadership

In the fall of 2006, I was stationed at Fort Bragg, which many call the "center of the Army universe." This description comes from the fact that two of the Army's most elite units, the 82nd Airborne Division and the Army Special Operations Command, are located there. I had just completed an adrenaline-filled airborne operation and was physically drained. Jumping out of airplanes was routine by this point in my career, but it still made me a little jumpy, pun intended. Many paratroopers reading this might call me a punk for admitting I was scared when jumping; after all, it was literally my job. I say any paratrooper who claims not to be afraid when they jump is lying.

It is okay; hell, it is *natural* to be afraid of jumping out of an airplane at an altitude of 1,250 feet above ground level, trusting a parachute to carry you safely to the earth below. Being *afraid* does not make you a *coward*. Let me clarify. The key distinction between fear and cowardice lies in this: a coward lets fear dictate their actions.

Unfortunately for me, during this airborne operation, the driver for a group of VIPs was injured and could not drive them back to the unit. I was "*voluntold*" to take the driver's place. "*Voluntold*" is a common Army term for when someone senior "asks" you to volunteer, but saying "no" is not an option. I was told to leave training early to drive the VIP group back to base. There is an unspoken military rule: "If you are the lowest-ranking person in the vehicle, keep your mouth shut and your ears open." So, that is what I did. I *listened*.

One of our physician assistants was among the passengers, discussing his transition from enlisted Special Forces medic to officer and physician assistant. It was a casual conversation about career paths and retirement plans. Then, out of nowhere, one of the colonels turned to me and asked, "How come you never got commissioned?"

I do not remember exactly what I said, but I gave him a flippant answer: "Who would want to be an officer?" I laughed it off, hoping to dodge the colonel's question because I did not want to admit that I did not think I was smart enough to be an officer, let alone pass the rigorous selection process. But the colonel did not let it go. He looked at me and said, "You would make a good officer. You should think about it." And that was that. It was a brief car ride. A casual remark. But it left a lasting impression on me. I am not sure why it stuck; it was never in my career plan, but it did all the same.

Later on, I started digging. I looked up the requirements for Officer Candidate School (OCS) and realized I *could* make it happen. With the proper support of a senior officer in my unit, I could have some of the eligibility requirements waived. One thing that could not be waived, however, was my General Technical (GT) score. If you are unfamiliar, the GT score is part of the Armed Services Vocational Aptitude Battery (ASVAB) test. The ASVAB is similar to the Scholastic Aptitude Test (SAT), but for military personnel. My GT score was 101; to qualify for OCS, I needed a 110.

I enrolled in an intensive prep course and retook the ASVAB. This time, I scored a 108, close, but not close enough. Army policy required a six-month wait before retaking the ASVAB. I did not have six months to spare. I was already older than most candidates; time was not on my side.

I started looking for options. I found a loophole that allowed me to retake the test a month later, and I scored a 109. Still one point short. I started second-guessing myself, wondering if maybe I was not cut out for this after all.

Perhaps I had already reached my ceiling and did not even know it. That kind of doubt can erode your motivation if you let it. I had been relatively successful in my career so far, and I was embarrassed that this was happening to me. I had told everyone I knew I had applied for OCS, and having to go back to them and admit I was not going to make it was a hard pill to swallow. I had to dig deep and remember my purpose.

I applied for a waiver because I was *so* close, and the Army responded. When I read the official answer, I froze. No waiver. No exceptions. It hit me like a gut punch. I was short by *one* point, one lousy point. Once again, I had to dig deep, recall my purpose, and stick to my plan. I had done everything right. I studied. I sacrificed my time. I found every loophole I could. I went back and retook the test *twice*. I was not asking for handouts. I was not looking for a shortcut, just a shot.

It was difficult *not to* despair. But that one point stood between me and everything I had been working toward. And in that moment, it did not feel like a point on a test. It felt like a rejection. It felt like a door slammed in my face by the very system I had poured myself into for *years*. It was not just disappointment; it was heartbreak. When you are *that* close, it does not feel like a professional failure; it feels personal.

I remember walking into my commander's promotion ceremony, carrying that weight. I did not even want to say the words out loud, but I owed him honesty. "Sir," I said quietly, "I probably will not attend OCS." He looked at me for a moment, calm and steady. "Get on my calendar," he said, "and come see me. We will figure something out."

It stopped me cold. Someone of his stature knew me and believed in me. He had seen me fight for others. He had seen me push through the impossible. It lit a fire. I had to dig deep, one last time, and remember my purpose. I left the ceremony feeling energized and determined to meet the challenge. In that strange collision of defeat and defiance, something inside me solidified. I realized I was not done yet. I returned to work, dug deeper than ever, and found another loophole, thanks to my Army brother Michael Bell, whom I had, under unusual circumstances, also convinced to go to OCS months earlier. Following

Michael's advice, I was *finally* commissioned as a Military Police officer in 2008, not because the road was easy, but because I refused to let that one point define the story. Time after time, knockdown after knockdown, I remembered my purpose.

In hindsight, that single point became one of the most defining moments of my leadership journey, not because it threatened to stop me, but because it tested me in a way other numbers never could. It is easy to romanticize resilience and talk about grit as though it is a noble badge. Real resilience is quiet. It is lonely. It shows up in the hours no one sees, when your confidence is shaken, your back is against the wall, and all you have left is the belief that this *still* matters. That *you* still matter.

That moment taught me something fundamental: leadership is not earned in comfort; it is forged in friction. Not just the friction of external challenges, but also the internal battle between giving up and pressing on, between silence and action, between almost and absolutely. As a leader, I have come to realize that the people worth following are not those who have traveled the smoothest road. They are the ones who have fallen and bled on their road, gotten up, and kept going despite the pain and heartbreak. They have heard "No" and tried again. They have stood at the door marked "No Admittance" and knocked anyway. They did not fight for a mere seat at the table; they fought for a voice in the boardroom. Above all, they have carried the weight of their disappointments with grace, turning them into fuel for their ongoing journey.

So it was never really about one point. It was about proving to myself, more than to anyone else, that I would not let someone else write the end of my story. Leadership is, at its core, all about refusing to quit, even when quitting feels justified.

If you are sitting there wondering whether you are cut out for leadership, remember this: it is not always the smartest, the strongest, or the most connected who rise. Sometimes it is just the most stubborn. Those who keep pushing, keep believing, and keep showing up. Do not wait for someone to open the door for you. Find the key, or build a new door.

Chapter Four

Why Real Leadership Has Nothing to Do with Position

I believe people will always follow competent leadership, regardless of title. I have held many ranks over my thirty-three-year career in uniform. From private to field-grade officer, I have held titles that came with power, responsibility, influence, and authority. Here is the truth: Titles do not necessarily make people follow you. Respect does. Trust does. Who you are when nobody is looking defines who you are.

I also firmly believe that people will always follow competent leadership, regardless of title. And I have seen it firsthand, time and time again.

During my time in the military, many of my units had what we called *Motorpool Monday*. It was the day we performed maintenance on our vehicles and cleaned the motor pool. Most of the time, leadership had a meeting beforehand to plan the week. Sometimes, they would show up late to *Motorpool* on Mondays.

Here is where it got interesting: We never waited for them. Without fail, someone would call out, not in a demeaning way, but clearly and confidently, "Let us get started!" People would fall in. It did not matter who said it. Nobody questioned their title. There was little complaining. We just lined up and got to work, picking up trash, checking equipment, performing maintenance checks, and getting the vehicles lined up perfectly.

Now, I already know what some of you are thinking: *"Greg, that type of situation only happens in the military."*

But let me stop you right there, because that belief could not be further from the truth. Leadership does not belong to a uniform, a rank, or a title. It belongs to people who are willing to step forward when everyone else steps back. Occasionally, the most powerful leadership moments appear in the most ordinary places.

Let me introduce you to my friend and brother, Louis, though if you really know him, you call him **Louie Lou**. Louie grew up in South Carolina and joined the Army at a relatively young age. He has one of those rare qualities that cannot be taught; you simply recognize it. When Louie talks, people listen, not because he demands it, but because he projects presence, confidence, and authenticity.

One afternoon, Louie and a few of his buddies were on a road trip and stopped for gas and food. It was lunchtime along a major highway, a combination that is guaranteed to create chaos in a fast-food restaurant.

Chaos is precisely what they encountered.

Orders were backed up, customers were irritated, and the young manager looked like he was moments from tossing his headset into the fry grease and walking out the back door. You could *feel* the stress in the room.

This is where most people shrug and say, "Not my problem."

However, not Louie.

He made his way up to the counter, not aggressively or arrogantly, but with calm, confident energy. He said to the manager with a smile, "Looks like things are a little wild. Want a hand?"

Now, here is the moment that matters: The manager said yes.

In saying yes, he broke every corporate policy in the handbook. No HR screening, no food safety certification, no paperwork, no training, nothing. But right then, he did not care. His most immediate problem was not compliance; it was angry customers and a mountain of backed-up orders.

In that moment, leadership mattered more than procedure.

Louie did not just help; he mobilized. He organized his friends like a seasoned squad leader. One handled the line, one communicated with customers, one supported the overwhelmed staff, and Louie transformed the chaos into order.

Within minutes, the energy shifted. People who had walked in frustrated were now laughing. Employees who were drowning in stress were following his direction with relief. Customers who had no idea who he was trusted him, because confidence is contagious.

Leadership did not come from authority. It did not come from a name tag. It did not come from permission. It came from action.

Why? Because people will follow someone competent and decisive, even if that person outranks no one. That is real leadership. And that is the question I want to ask you: If your title disappeared today, if no one had to follow your orders, would they still follow *you*?

I have watched people with high ranks get ignored. I have also watched junior soldiers and civilian interns step up and lead. Not because they had a title, but because they had influence. Leadership is not about being the boss; it is about being the person others believe in.

Let us ask the fundamental question: Would your subordinates follow you if you no longer had a title? That brings us to the heart of leadership, the difference between position and influence. One gives you authority on paper; the other earns you respect in reality.

Define the Difference: Position vs. Influence

Let me explain something that separates authentic leadership from the illusion of it. When you lead from position, your rank, title, or where your name sits on the org chart, people might do what you ask. But often, it is only because they *have* to. That kind of leadership gets you the bare minimum: surface-level respect, checked boxes, and a lot of head nodding in meetings. But the minute you walk out of the room, so does your influence.

Now, contrast that with leadership based on *influence*. Influence does not require a title; it is built over time through consistent action, character, and follow-through. People follow you not because they are obligated to, but because they trust you. Because they have seen you lead from the front, not bark from the back.

Every great organization I have ever served in had people like this. Some were junior enlisted. Some did not have any official authority at all. However, when things became brutal or chaotic, people instinctively looked to them for direction, not because they were told to, but because they knew those individuals could be trusted.

I have seen a corporal carry more weight in a squad than the staff sergeant or the squad leader. I have also seen a civilian analyst in a planning room command more respect than a full-bird colonel. Why? Because when they spoke, people *listened*. When they acted, people followed.

That is the difference.

A position might grant you authority, but influence earns you respect, and respect lasts longer than any title. If your title were to disappear tomorrow, what would remain? Would people still come to you for guidance, support, and direction? Would they follow your lead when no one was making them?

If the answer is no, it is time to stop leading from the nameplate on your door and start leading from *who you are*. Let me provide a real-world example.

I want to tell you about my friend Melanie.

She and I worked together in a warehouse after we both transitioned out of the military. It was a fast-paced, high-pressure environment, with boxes moving, deadlines looming, and people constantly on the go. In a place like that, leadership stands out quickly, and so does the lack of it.

Melanie is a retired Army master sergeant, which is impressive enough. But what truly sets her apart is how she earned those stripes. She rose in a career field where very few make it that far. It took consistency, resilience, and the kind of respect you only get from years of doing the hard things the right way.

Now, here is the interesting part: Melanie is not a logistics expert. Her background is not in supply chain or warehouse operations. But if you spent five minutes in that building, you would swear she ran the place. Everyone, from leadership to associates and supervisors, knew who she was. More importantly, they *trusted* her. She never led with her résumé or her position. She led by *taking care of people*.

She knew the associates by name. She asked questions without ego and took classes to further her education. She pitched in when no one was looking. I have seen her sweep up after a shift that was not even hers. I have watched her quietly bring out water and Gatorade to the dock on blistering days. And when someone was having a tough day, she was the one they would confide in, whether they reported to her or not. She did not need a title to lead; she just showed up and served, over and over again.

I picked an associate to join my team solely on her word. One of the younger associates said something I will never forget: "I do not even work on her team, but if she asked me to do something, I would do it without hesitation." That is the kind of leadership you cannot fake. That is *influence*.

Melanie reminded me of a truth we all need to remember: real leadership does not require permission. It does not need a job title. It requires someone willing to lead with humility, consistency, and a genuine heart.

To determine whether people will follow you without a title, observe how they treat you when you are *not* in charge. That is where authentic leadership begins. So how do you build that kind of influence when you do not have a title? The answer is not complicated, but it does require consistency. Here are five practices I have seen separate real leaders from those who only look like leaders on the org chart.

How to Build Leadership Without a Title

You do not need a promotion to start leading. You do not need a plaque on your desk or your name at the top of the roster. What you *do* need are a few core habits and traits that people naturally respond to. If you want people to follow you, whether you have a title or not, focus on these five steps.

1. Be Consistent

People will follow what is predictable, not just what is impressive. If your team ever has to guess what kind of leadership style they are going to see on a given day, you are off the leadership mark. Are you consistent in your attitude, your effort, your communication? The worst kind of leader is the one who is up one day and down the next, full of motivation on Monday, silent and distant by Thursday.

Consistency builds trust. It shows your people they can count on you, especially when things get hard. You do not need to be perfect; you just need to show up the same way, day in and day out. Think of the people you trust most in your life. Odds are they are the ones who are steady. Not flashy. Not dramatic. Just dependable. You do not need a title to lead.

1. Communicate Clearly and Often

You do not have to have all the answers. But people should never have to guess where you stand. Be honest about what you know, what you do not, and what you are doing to figure it out. When things go wrong, do not hide, talk about it. When things go right, share credit freely. Leaders without titles do not have the luxury of hiding behind hierarchy. Their communication *is* their credibility. I always told my teams, "If I do not tell you what is going on, you will write your own story, and it probably will not be a good one." People do not expect perfection; they expect honesty. You do not need a title to lead.

1. Earn Trust Through Action

Talk is cheap. Everyone knows that. So, if you say you will do something, do it. If you set a standard, meet it first. If you expect others to stay late, demonstrate that you are willing to do the same. This is not about playing martyr; it is about being the kind of leader who does not ask for anything they would not do themselves.

When people see that your actions match your words, your influence skyrockets, and when they do not, you lose more than just respect; you lose the opportunity to lead. Let me tell you how I learned that trust does not come from your title. It comes from what you do when no one is watching, and, especially, from what you do *every day*.

When I took over a particular organization, one of the first things I noticed was a disconnect between leadership and the team. The soldiers were not disrespectful, but they were not exactly motivated, either. There was a clear sense that they had been let down. I later found out why. My predecessor rarely attended morning PT.

Now, that might sound minor to some, but in the military, PT is not just about physical fitness; it is about discipline, accountability, and shared struggle. When your commander does not show up, the message is loud and clear: "This is not important."

So I did the opposite.

I showed up every morning, in the cold, in the heat, rain or shine. Not just to check the box, but to *do the work* alongside them. I ran the runs, did the push-ups, and took the same PT tests they did. I showed up for the PT test, and a soldier approached me and said, "What are you doing here?"

"I am here for the PT test."

The young sergeant said the leadership usually took the PT test with just a group of leaders. If I said something mattered, like motor pool maintenance, leader meetings, or squad-level training, I was there. Not just occasionally, every time.

No big speeches, no "look-at-me" moments, just consistency. At first, people said little; they had seen leaders come and go. They were waiting to see if I meant what I said. But slowly, I started to notice a shift. Soldiers began arriving a little earlier. Their effort led to small improvements, tighter formations, fewer complaints. One Noncommissioned Officer (NCO) came up to me after a run and said, "Sir, no offense, but we did not expect to see you out here this much. The fact that you are, it means something." That is all I needed to hear.

Because here is the thing: you do not earn trust with one big moment. You earn it in the little moments most people overlook, by showing up consistently without needing applause. Your presence sends a message. And when your actions match your words, people do not just follow orders; they follow *you*. You do not need a title to lead.

1. Practice Empathy Without Excuses

Being empathetic does not mean lowering standards. It means seeing people, hearing them, and understanding that behind every uniform, ID badge, or time card lies a human being with real-life struggles. A great leader can look someone in the eye and say, "I expect the best from you," and also, "How can I support you while you give your best?"

The leaders I respected most in my career were tough but fair. They did not let us off the hook, but they also did not treat us like robots. They saw us, and that made us want to give them everything we had. Let me tell you a story that still sticks with me.

I had a soldier, an MP, who got a DUI. Now, if you are in the Military Police Corps, that kind of mistake is not just personal, it is professional. It is like a bank teller stealing money. You are supposed to be the standard-bearer. So when you break that standard, the punishment typically hits even harder.

Most people expected me to hammer him. Frankly, that would have been the easiest thing to do. But I did not forget where I came from.

Early in my career, I found myself in a very similar situation. I had made a grave mistake, one that could have ended everything I had worked for. But instead of ending my career, my battalion commander pulled me aside. He looked me in the eye and said, "You messed up. There is no sugarcoating it. But this does not have to define you."

He showed me empathy. He held me accountable, but he also gave me space and support to learn, grow, and come back stronger. That moment changed the trajectory of my entire career, and I never forgot it. So when that soldier came into my office, I remembered how it felt to be on the other side of the desk.

I asked him, "What is going on?"

The uniform came off, so to speak. He told me everything: family issues, stress, drinking to cope, and the weight he was carrying that no one saw. He was not dodging responsibility. He owned it. But he was also unraveling, and no one had ever asked why. I did not let him off the hook. He still faced the disciplinary process. However, I also made sure he had support. I connected him with resources, stayed engaged with his progress, and made sure he knew he had not been written off.

I told him what had once been said to me: "You messed up, but this does not have to be the end of your story."

And it was not.

He worked his way back, earned back the rank and trust he had lost, and recommitted to the team. Eventually, he became one of the leaders others looked up to. Years later, I ran into him, and he thanked me, not for going easy on him, but for seeing the man behind the mistake.

That is empathy. That is leadership, not making excuses, but showing up for someone the same way you would hope someone would show up for you.

1. Serve First, Lead Second

Servant leadership is not a slogan, it is a way of life, and one of the most powerful ways to build real, lasting influence. We have all seen the opposite, leaders who want the perks of power without the responsibility. Leaders who show up late, leave early, and let others do the heavy lifting. They hide behind their title, but the truth is that nobody wants to follow them when it counts.

People do not follow titles; they follow character. Serving first means you are willing to do the hard, humble things without expecting applause. It means asking, "How can I help?" instead of "What do I get?" It is showing up early, staying late, and doing the thankless jobs, because your people are watching. And whether you realize it or not, they are deciding whether you are worth following.

When you show your team that no task is beneath you, they will believe that no mission is beyond them. When you support them, they will go above and beyond to help you. And when they know you are not just in it for the position or the praise, they will run through walls for you.

I have been in rooms where the most respected person was not the highest-ranking one; it was the person who always showed up. The one who asked about your family, who carried the gear without being asked, and who stayed behind to clean up. No fanfare, just a quiet, consistent presence. That is what real leadership looks like, not loud, not flashy, just dependable, humble, and invested in the success of others.

Why It Matters

Serving first sets the tone; it fosters a culture where people feel supported and valued. It sends a message that the leader is not above the team; they are part of it.

And when times get tough, people do not rally around a title; they rally around a person who has *earned* their trust through action.

People will work hard for a paycheck, but they will *go to war* for a leader they know has their back. If you want to build a high-performing, loyal team that not only meets the standard but raises it, you need to model what it looks like to put others first.

- Are you willing to do the job you are asking others to do?

- Do you take responsibility when things go wrong?

- Do you give credit when things go right?

Serve first. Lead second.

Bottom Line: Influence Is Built, Not Given

You do not earn followership with a title; you earn it through consistency, credibility, compassion, and action. You build it one interaction at a time, one moment when you chose to lead, not because you *had* to but because it was the right thing to do.

Some of the most outstanding leaders I have worked with did not sit in the corner office, they were on the front lines. They did not carry fancy titles, but they carried trust, effort, and integrity, and people followed them.

So the question remains: Would people follow you if you did not have a title? If the answer is yes, keep going. If the answer is no, the good news is that leadership is a choice, you can start leading right now.

Hard Reveals Who You Are

One of the biggest challenges I faced in my life and my military career was attending Ranger School. However, I would like to clarify what is probably one of the biggest misunderstandings about the word "Ranger." I would like to explain the distinction between the 75th Ranger Regiment and United States Army Ranger School. The 75th Ranger Regiment and Ranger School are separate entities, although they share the word "Ranger."

The 75th Ranger Regiment is an elite special operations unit in the U.S. Army. It is a permanent assignment. Rangers in the 75th are full-time, highly trained soldiers who specialize in direct-action raids, airfield seizures, special reconnaissance, and high-value-target missions. They live, train, and deploy as part of a tight-knit combat unit. Being selected for the 75th Regiment requires passing a challenging selection course called the Ranger Assessment and Selection Program (RASP) before serving in the regiment.

Ranger School, on the other hand, is a leadership training program open to almost everyone in the Army, although few will volunteer for the rigorous course. If I had a quarter for every time I heard someone say, "I would go to Ranger School but (insert excuse here). Otherwise, I would be a millionaire." An old sergeant once told me, "There are those with *tabs*, and there are those with *excuses.*"

At the United States Army Ranger School, students are pushed far beyond their comfort zones to reveal who they will be as leaders. With only two to four hours of sleep a night and barely enough food to sustain them, Ranger students are constantly wet, cold, hungry, and exhausted. Their rucksacks (large backpacks) usually weigh between sixty-five and ninety pounds but often tip the scales at more than 100 pounds once weapons, radios, and other gear are added. They carry that burden across mountains, swamps, and miles of punishing terrain. Every step grinds them physically and mentally, but that is the whole point. Ranger School is not supposed to be easy, and it is not about who can run the fastest or carry the heaviest ruck; it is about who can still think, make decisions, and lead others when they are operating on empty.

When I went to Ranger School, it was a grueling seventy-two-day course that had four phases: Darby at Fort. Benning, Georgia; Desert at Fort Bliss, Texas; Mountains at Dahlonega, Georgia; and Swamps at Eglin Air Force Base, Florida, but the Army later eliminated the Desert phase. I am not sure why, but I assume it is because of the cost of flying all the students to the desert when the other phases are only a bus ride away. The inclusion of the Desert Phase made my class the "last hard Ranger class." Every Ranger School graduate is now laughing as if it were an inside joke. I apologize to those who may not understand. The course is open to soldiers from across the Army, and occasionally other services, who wish to demonstrate their skills in small-unit tactics and leadership under extreme stress. All graduates of Ranger School earn the Ranger tab, but not everyone automatically joins the 75th Ranger Regiment.

Leadership in any environment is not about giving commands; it is about inspiring trust in others when everything inside you really wants to quit. It is about prioritizing your mission and your team over your comfort and discovering the strength you did not know you had. Ranger School strips away all the advantages and distractions of everyday life, forcing students to confront their weaknesses and rise above them. If they do not, they will be sent home.

A perfect example of the candidate who will be sent home is the Spotlight Ranger. Spotlight Rangers never make it to graduation, and I daresay that they are rarely remembered as good leaders, no matter where they end up. A Spotlight Ranger is the Ranger student who only does the right thing when the instructors are watching. This is a term that I use to this day. Ranger instructors are not just looking for someone who can perform under the pressure of the spotlight; they are searching for the kind of leader who steps up and drives the team forward, whether eyes are on them or not. How many of you have that one coworker, usually the laziest person on the team, who magically transforms the second the boss walks in? One minute, they are doing nothing. Next, they are acting like the hardest-working, most dedicated employee. It is all a show, and everyone knows it. Here is a test: If I asked everyone in your office who the spotlight is or who the "deadbeat" on the team is, I can almost guarantee that everyone on the team would pick the same person. Let us hope that person is not you!

Ranger School instructors are as focused on how you *lead* as they are on how you *follow*. In Ranger School, leadership positions rotate constantly. Some days, you are giving orders; other days, you are taking them. Either way, how you act when you are not in charge matters just as much and sometimes even more than when you are in charge. Authentic leadership is not about playing to the crowd. It *is* about consistency, character, and showing up strong even when nobody is clapping.

I was a medic when I went to Ranger School. I did not *need* to go to this school to succeed in my Army career, but I was not about to let that stop me.

Before I joined the Army, my uncle, Alejandro Archbold, gave me some advice that will stick with me for life: "Take advantage of every opportunity the Army offers you. Never turn down a school." What I took from that was to never miss an opportunity to challenge yourself.

I arrived at Fort Benning, Georgia, in 1992. Fort Benning is home to the Infantry, and I knew I was standing at the gateway to some of the most challenging schools the Army had to offer. Even though I was a medic, not an infantryman, I wanted in. I refused to let an opportunity such as Ranger School pass me by. People kept asking, "Why would a medic go to Ranger School? You do not need it for promotion." They did not get it. I was not chasing a checkbox; I was chasing excellence.

At that time, you could walk around Fort Benning for hours and never see an African American soldier wearing a Ranger tab. Then, one day, while providing medical support for Ranger School, I saw two medics with Ranger tabs: Staff Sergeant William P. Franklin and Sergeant Kevin Tillman. I was starstruck. Both Rangers would go on to have legendary careers. Seeing them lit a fire inside me. I submitted my 4187 request form and made it my mission to become one of the few African Americans with a Ranger tab. Being a medic was simply more icing on the cake.

Of the 120 soldiers in my class, ten were African American. There were always whispers that the Ranger School did not want to pass Black soldiers. I did not care about rumors; I cared about proving that I belonged. I pushed myself through every single phase.

The road to success, however, is not a straight line. In the third phase, things came crashing down. Instead of being recycled like most students who did not meet the standard, I was told I was going home, with no second chance and no explanation. It was a crushing blow. I rode the "no-go bus" back to Fort Benning with my head low and my heart heavy. Most of the soldiers on that bus were African American. Was it racism? I thought it was, but I had no evidence to support my theory. In situations like this, I learned early in life that it is not about what you *believe* but what you can *prove*.

Regardless, I was not about to let one failure define me. Often, life is not fair; if you want to succeed, you need to get over it and get better. Therefore, I went back to work. I worked harder. I worked *smarter*. Two years later, while stationed in Vicenza, Italy, I fought for a second chance. This time, I had to attend another pre-Ranger program and compete for a slot against some of the best foot soldiers in my battalion. Sergeant First Class Tony Mendoza, the scout platoon sergeant, supervised the pre-Ranger program. I shared my story with him, and he fully supported my decision to enter the program. I earned a Ranger School slot. This time, I graduated with Class 10-94, just in time to witness the birth of my son, Javier.

Here is the bottom line: Never quit on yourself. No matter what the odds look like. No matter how many people doubt you. Quitting guarantees one thing: you *lose*. Ranger School did not only test my endurance; it revealed principles that I carried with me for the rest of my Army career and will continue to use for the rest of my life. The lessons were not written in a manual; they were carved into me, mile after mile, ruck after ruck, setback after setback. The truth is, you do not have to wear a uniform or carry a rifle to apply these lessons to your life. They are universal. They are about grit, leadership, and the refusal to quit when everything in you screams, "*Stop!*"

Here are the takeaways that Ranger School branded into me:

1: Do Not Wait for Permission to Be Great

No one sent me a personal invitation to attend Ranger School. I went because I chose to go. I saw a challenge, and I ran toward it. I did not wait for validation, permission, or the perfect moment. The time is always *now*. Too often, we hold back. Excellence does not ask for approval; it demands courage, decisiveness, and self-leadership. I knew this would be a physical and mental test of my leadership, and I was up for the challenge. If you want to be a leader, start by leading yourself, not through the straight line of ease, but through the corridors of struggle. You must look for opportunities and take them when they present themselves. You do not need external validation or a checklist to pursue excellence. Your excellence does not need a permission slip; it requires your unwavering commitment.

2: Representation Matters

The first time I saw Staff Sergeant Franklin and Sergeant Tillman, something shifted in me. They were not just excellent at what they did; *they looked like me*. That simple fact mattered more than I realized at the time. It did not take a motivational speech or a written policy to inspire me; it took two professionals standing confidently in spaces I had not yet imagined for myself.

Role models matter. *Visibility* matters. Sometimes, seeing someone who looks like you, who comes from the same place you do, and who thrives in a place of excellence unlocks a new level of belief. It silently encourages you to dream bigger, to reach higher, and to know that the ceiling is not where others have placed it. The ceiling is as high as you allow it to be, and only you can decide to punch through it. Greatness is not reserved for a few; it is available to all who dare to see themselves in it.

Let your presence become someone else's permission.

3: Lead When Nobody Is Watching

"Spotlight Rangers" might be able to fool a few people for a little while, but spotlights never last. They may capture your attention, but they will never garner your respect. Do not be a spotlight! Great leaders do not perform for credit; they grind for impact. Leadership is conducted when the lights are off, when nobody is watching, and when quitting would be easier than continuing. Leadership is not about being seen; it is about being dependable in the eyes of those who need you.

4: Hard Reveals Who You Are

Ranger School strips you of comfort, status, and ego; what is left is the real you. And that is the point. Authentic leadership is not revealed in perfect conditions; it is forged in the furnace of hardship. Your leadership style should be the same whether the waters are as smooth as glass or crashing around you in stormy waves. In most turbulent times, your team does not just need your direction; they need your calm. Your reaction becomes their blueprint.

5: Failure Is Not Final—Unless You Quit

Being dropped from the program after nearly earning Ranger of the Cycle, which is another term for MVP of my class, could have been the end of my story, but I decided to write a better one instead. I did not allow rejection to break me. Do not allow it to break you. The no-go bus was not the end of the line. It was a detour on the journey. When I returned to Ranger School for a second time, it was to finish what I started. Fail*ing* is part of the process; you only become a fail*ure* if you quit.

Ultimately, Ranger School was not about a tab on my shoulder. It was about proving to myself that I could endure, rise above, and lead when quitting was easier. The same is true for you. Your obstacles may not look like mountains, swamps, or 100-pound rucks, but the principle is the same: hard reveals who you are.

Do not wait for permission to be great. Do not quit when failure shows up. Do not live for the spotlight. Lead with consistency. Represent possibility for others. Above all, remember your most challenging moments are not there to break you; they are there to build you.

Quitting guarantees that you lose, but persistence guarantees that you grow.

Chapter Five

The Leadership Trifecta

Purpose, Presence, and Passion

L eadership does not come with a certification exam. But it is one of the most critical roles in any organization, and yet anyone can be promoted into it without a single hour of formal training.

Perhaps that is why 31% of new hires in a recent survey *reported not trusting their leadership*. Maybe that is why many of you have, or will in the future, use phrases like "Why is this guy in charge?" and "She does not get us," and "I could do a better job myself." Sound familiar?

The truth is, most leaders are figuring it out as they go, learning by watching those who came before them. Leadership is like parenting: there is no manual, only responsibility, and you learn by doing. Just like parenting, leadership habits are passed down. Some are good, and some should have been left behind years ago. I believe we can do better. I think we *must* do better.

Since 1990, I have led soldiers through war zones, humanitarian missions, and leadership trenches that cannot be taught in a classroom. What I discovered is that great leaders consistently operate from three foundational elements, what I call the **Three Ps: Purpose. Presence. Passion.**

Purpose

I remember the day clearly. We were in Afghanistan, operating out of Forward Operating Base (FOB) Shank, but most of our dealings were in a small town called Charkh, surrounded by dust and uncertainty. We had just returned from spending the night at one of the checkpoints, called the Dabari Bridge. It was the kind that drains you, not just physically, but mentally as well. Everyone was worn out. But there was a kind of quiet that usually does not happen in a combat zone. Instead of heading back to FOB Shank to refit and get a much-needed shower and a few Rip-Its, we headed to the police station to meet with our Afghan partners for additional training.

Then the call came in.

One of our sister units had been hit by an IED. They needed immediate help, support, and security. We were the closest unit to assist, so we sprang into action. Even though we had trained for this kind of response, we never thought we would be called upon as the Quick Reaction Force (QRF). Not because we were not ready, but because, would the infantry ever call on the MPs? To me, it did not matter whether they called us because we were their only option or because they trusted we would come and help; our purpose would drive our actions.

I looked around at my team. My soldiers were already dirty, tired, and hungry. I saw it in their faces: fatigue, frustration, even doubt. We had a choice. And that is the moment every leader should prepare for, not the easy ones; those take care of themselves.

The ones that demand something from your people when they feel as if they have nothing left to give. And in that moment, I did not need a speech. I simply relied on **purpose**.

I did not have to remind them that we were not just soldiers but brothers and sisters whose purpose was calling. That our job was not over until every teammate came home. That if it were *us* out there, we would want someone to come to our aid. And if not us, then who?

They needed clarity. And purpose gave it to them. We rolled out minutes later, tired but focused. I watched the transformation as we prepared to move, with no hesitation and no dragging of boots. There was a fire in their eyes, and as a leader, I loved it.

Because purpose does that, it does not take the weight off your shoulders, but it gives you a reason to carry it. We got to the site and secured the area. Luckily, there were no wounded, but one of the vehicles was still on fire from the IED.

We stayed until the fire was out and the vehicle could be safely towed back to the FOB. No one said much during the ride back, but I could tell each one of them knew we just *lived* the reason we signed up in the first place.

That day was not just about tactics or mission success. It was about identity. About choosing service over comfort. Purpose over convenience. And that has stayed with me every day since.

Leadership is not about commanding people. It is about anchoring them. And in chaos, the strongest anchor you have is not your title, rank, or skill, it is your purpose. That day reminded me: *Purpose does not just guide you. It galvanizes everyone around you.*

What Does Purpose Look Like in Authentic Leadership?

Purpose is not a mission statement on the wall or some corporate philosophy. It is the reason behind the hard decisions. It is what keeps you grounded when things go sideways. Authentic leadership is not about doing the job; it is about understanding why the job matters.

I have seen leaders who were intelligent, charismatic, and capable, but their teams never followed them with conviction. Why? Because they lacked clarity of purpose. And if you are unclear about your "why," your people will be unclear about theirs.

On the other hand, I have followed leaders who were not the loudest or most polished, but they knew exactly why they led, and it was contagious. When a leader's purpose is genuine, it is evident in how they carry themselves, how they serve others, and how they navigate adversity.

Leadership with purpose creates teams that do not just comply; they commit. They buy in. They believe. So, how do you build that kind of culture? You do not just talk about purpose. You show it, you live it, and you reinforce it every day. There are ways to install purpose within your teams.

Explain the *Why* Behind the Work

One of the most prominent mistakes leaders make is assuming their people already know why what they are doing matters. That assumption leads to disconnection, and disconnected teams do not give their best effort, they provide the bare minimum.

Purpose does not live in job descriptions or mission statements. It lives in the *why* behind the work. As a leader, your job is to constantly connect the dots between what your people are doing today and the mission you are working toward together.

When I was in command, we had soldiers assigned to mundane tasks, such as inventory checks, motor pool maintenance, and data tracking. None of it looked heroic. But I never let them believe it did not matter. I would say, "When we deploy, lives will depend on whether this gear is accounted for. That spreadsheet you are filling out, that is not paperwork; that is preparation. That is how we lead before we even show up."

People will rise to the level of their understanding. If they understand how their work contributes to the mission, they will take ownership. However, if they feel as if they are just checking boxes, you will get compliance at best and burnout at worst.

How to Explain the *Why* Behind the Work

Tie tasks to impact in real time. Do not wait for a quarterly review or a town hall meeting; take action now. Every time you delegate or assign a task, take ten extra seconds to explain how it supports the bigger picture.

Use "mission moments." Start meetings by highlighting how someone's effort moved the team closer to the goal, no matter how small the task may seem.

Reinforce purpose in one-on-ones. Ask your team members how they see their roles contributing to the mission, and listen to their responses. It shows you care and reminds them that they matter.

Celebrate impact, not just performance. Be the subject-matter expert, and if you do not know, say so. Find the answers.

Many leaders believe that recognition means handing out awards or applauding at team meetings. That is surface-level. If you want to instill purpose, do not just celebrate who got the task done; celebrate how their effort *mattered*.

Purpose grows when people feel as though their work makes a difference. When they know they are not just checking boxes, they are advancing the mission. That is where real motivation lives.

In every high-performing unit I have led, I made it a priority to recognize subject-matter experts not just for being technically sound but for how their knowledge helped others win. That kind of recognition tells your team, "What you know and how you use it matters. You are not just here to work; you are here to lead."

When you recognize impact, you create clarity. You show the team what "right" looks like. You reinforce that purpose is not found in a title; it is found in how you apply your skills for the good of the team.

How to Incorporate the Celebration

Call out the *why* behind the win. Do not just say, "Thanks for doing that report." Say, "Your analysis helped the whole team pivot faster. Your insight made the mission easier for everyone."

Shine a light on expertise. When someone solves a problem because of deep knowledge or experience, highlight it. Purpose is reinforced when people see their strengths being used for something bigger than themselves.

Recognize unseen efforts. Sometimes, the most critical work is not flashy; it is quiet, consistent effort. Celebrate the people who keep the machine running. That is leadership too.

Let Your Example Set the Standard

You can talk about purpose all day long, put it on PowerPoints, preach it in meetings, but if your team does not *see* it in you, they will not *believe* it for themselves. Purpose is not a slogan or a mission statement; it is something you model every single day.

Your actions are the loudest message you send. When you lead with clarity, conviction, and consistency, when you show up as if what you do matters, your team does not just follow your instructions; they follow your example.

No one embodied this more than Dr. Martin Luther King Jr. Dr. King did not just speak about justice, he lived it. He marched in the streets. He was arrested for what he believed to be true. He faced threats, violence, and exhaustion. Still, he kept going. Why? Because his purpose was not just words, it was his way of life. Think about it: he gave the "I Have a Dream" speech in front of hundreds of thousands because he *lived* that dream, even when the world around him did not.

That is what made his leadership so powerful: his actions *validated* his message. Do your leadership actions validate your message? Do you want to lead with purpose? Then live it when it is hard. Speak it when it is unpopular. Walk it even when you are walking alone. When your team sees you live what you preach, they will not just respect your leadership, they will rise because of it. I have created four steps on how to incorporate this regularly.

Connect with People, Not Just Projects

What it looks like: Spend the first fifteen minutes walking the floor (either in person or virtually). Check in not to manage performance but to ask how people are doing. Celebrate progress. Listen for struggles. Show that you care about *who* they are, not just *what* they do.

Purpose Modeled: This shows your team that people come before productivity. It reminds them that your leadership is grounded in purpose, not just profit. When people feel seen, they engage with deeper meaning.

Demonstrate Integrity in Tough Decisions

What it looks like: When you are under pressure, facing budget cuts, missed targets, or challenging conversations, do not take the easy way out. Be transparent. Be honest. Own your mistakes and stand by your values, even when it costs you something.

Purpose Modeled: Your people will take notice. If you do not compromise your principles, they will be more likely to stand on theirs. Integrity becomes contagious when it is modeled, not mandated.

Recognize Purpose-Driven Behavior in Others

What it looks like: Instead of only praising performance or output, start celebrating when team members act with purpose, whether it is someone going the extra mile for a client, mentoring a peer, or making a values-driven choice. Tell the story. Shine a light on it.

Purpose Modeled: You are telling the team that *how* they do the work matters as much as *what* they do. Purpose is not just a leadership expectation; it becomes a cultural norm. If your people can see purpose in *how* you lead, they will discover purpose in *why* they follow.

Presence: Be Where It Matters

I have served under leaders who were technically sound but lacked presence. Their teams followed instructions, but they did not buy in. Then there were leaders like Colonel Patrick D. Frank, under whom I served during my final tour with the 3rd Brigade, 10th Mountain Division.

One of his core beliefs? Leaders needed to be both physically fit and physically present. He did not send that message from behind a desk or post it in an operations order. He showed up. On time. In uniform. In formation.

Imagine this: nearly 4,000 soldiers in a brigade, and there he was, running, rucking, sweating, showing up where the work was happening.

One morning, my platoon and I were getting ready for a circuit workout followed by a short run. As we were stretching, COL Frank walked up. "What is on the schedule this morning?" he asked. When I told him, he replied without hesitation, "Let us get after it." And he jumped right in.

The soldiers loved it. And this was not a one-time stunt. It was who he was. This was a leader who did not just preach standards, he lived them. And because he did, we followed him with a level of commitment that went beyond compliance.

Importance of Presence

Presence is not about being seen; it is about being involved. It is how you prove the mission matters, that your people matter, that you are not too busy or too important to show up.

First, whether it is physical or virtual, presence is the foundation of trust. Your role as a leader is to inspire others to achieve the mission *and* to cultivate the next generation of leaders. But how do you develop trust if your team never sees you? Your presence matters.

Second, presence dictates importance. People show up for what they value. You attend weddings because your presence matters. You are there when your kid hits the stage or the field because you must. So when your organization is under pressure or moving fast, your presence sends a clear message: "This matters. You matter."

Real presence creates a connection. It builds credibility you cannot fake. And it sets a standard others will rise to meet.

I have served under leaders who could recite doctrine but did not know their people. Their teams did what was required, nothing more. But with COL Frank, there was buy-in. Soldiers did not just respect him because of his rank; they respected him because he was with them.

What Happens When Presence Is Missing?

Think back to elementary school. What happened when the teacher stepped out of the room? Chaos. Why? Because presence equals stability. The same applies to any organization. When a leader disappears, trust erodes and focus drifts.

Leadership presence offers something more profound than mere visibility; it signals a genuine commitment. It tells your people, "I am not just invested in the outcome; I am invested in *you*."

This Is Not Just a Military Lesson

In today's hybrid and remote work environments, presence does not always mean physical proximity, but it always requires intentional engagement.

Consider these situations to test if there is intentional engagement and presence.

- A Chief Executive Officer (CEO) who regularly walks the factory floor to listen to line workers, not just managers.

- A team leader who joins customer service calls to understand real-time pain points.

- A remote manager who checks in beyond project status, asking, "How are you doing?"

These small acts of intentional presence build loyalty and increase productivity. When people feel seen, they show up with more heart.

The Leadership Litmus Test

Leadership is not about being in charge, it is about being available. It is about being present, again and again, even when it is inconvenient, especially when it is difficult. So here is the challenge:

- Where are you showing up?

- What are you showing up for?

- And when you are present, what message are you sending?

Because, at the end of the day, presence is not optional; it is leadership.

Passion: Lead with Heart or Do Not Lead at All

Passion is defined as a barely controllable emotion. In leadership, it is not just about energy; it is about belief. Passion is the electricity that not only sparks the movement of leadership and the organization but also serves as the current that keeps everything moving forward. So, how do you get passion? And, more importantly, how do you maintain it? Simply put, you must have a strong *why*.

You cannot get a group of people excited about the mission if the leader is not enthusiastic about it. Your team does not need noise; they need conviction. Let me say this clearly: if the leader does not care, no one else will.

That does not mean you have to be loud or flashy. It means you must believe in the work, believe in your people, and believe in your purpose. Your passion is the heartbeat of the organization. If you go cold, everything else flatlines.

Simon Sinek said it best: "It is not about being in charge; it is about taking care of the people in your charge."

The Power of Passion in Action

I once led a small team supporting a field training exercise. It was not anything out of the ordinary; it was simply an essential part of preparing the unit for deployment. Then came the ripple.

One of my junior leaders made a poor decision that caused significant backlash. The squad leader was removed from the field, and the team was left without supervision, feeling embarrassed and isolated, and everyone in the unit knew it. Morale bottomed out.

The next morning, I headed out to their site. I was going to be the leader they needed, not reactive, not angry, but steady and present. I reminded them that this was not their fault. I told them it was time to get back to business and show the company what we were made of.

I stayed with them. I did not take over. I helped where I could. I stayed positive, even when the weather turned bad and criticism began to roll in, because I knew that if I showed up with passion, they would follow. And they did.

Passion is Contagious

I once saw a soldier struggling during her two-mile run for the Army fitness test. I did not know what lap she was on, but I could tell she needed encouragement. So I jumped on the track and ran with her, motivating her the whole way.

About a week later, another soldier came up to me and said, "Sir, I heard about what you did. I have never had a commander like that." That hit me in the heart. That is what passion does; it leaves a mark. It builds belief. It appears when nobody else can.

Reignite Your Passion

You can rebuild or rekindle your passion by understanding your *why*.

- Why did I want to lead in the first place?

- Who am I doing this for?

- What does my team deserve from me?

You do not need to be perfect, but you do need to show up with belief. When your people see that you believe, they will too.

Lead with conviction, or do not lead at all. Which P do you need to strengthen?

Chapter Six

The Power of Silence

Why Great Leaders Listen First

The greatest leaders do not talk first; they listen first. I was new to Fort Bragg, young, energetic, and eager to join the storied ranks of the 82nd Airborne Division, like so many paratroopers before me. During in-processing, I met the medics in my new platoon and, more notably, my new squad leader. He was loud and assertive, making it clear from the outset that he had no tolerance for laziness. Without much ceremony, he handed me over to my sponsor to finish in-processing. Two details from that first meeting still stand out: I did not get a single word in, and I was wearing my physical training uniform, just a T-shirt and shorts, which revealed none of my experience or qualifications.

A few days later, as I wrapped up paperwork across the base, my sponsor suggested that I meet the battalion's command sergeant major. We ran into him on the stairs as we left the aid station. He greeted me warmly and asked about my transition, my goals, and everything I had accomplished so far. We spoke for about fifteen minutes until my squad leader showed up out of nowhere.

He jumped into the conversation, eager to claim he was "mentoring me," and insisted that I aim for Jumpmaster and Ranger School to succeed in the 82nd, none of which he had accomplished himself. After a few moments, the CSM interrupted him.

"You do know he has already completed both of those schools?"

The silence that followed was unforgettable. The CSM shook our hands and moved on, but that moment said everything. My squad leader and I never quite got along after that. He had not learned a thing. He continued talking as if only his opinion mattered, silencing anyone in the squad who dared speak up with the tired phrase, "If the Army wanted you to have an opinion, it would have issued you one."

Leadership is not about having all the answers; it is about knowing where to find them. Too often, leaders assume that their ideas are most important, unintentionally silencing the voices that could drive real progress. Outstanding leadership begins with listening.

That does not mean leaders should not lead or make tough calls. It means that effective leaders understand the limits of their own perspective. They know that the closer someone is to the work, the greater their insight into improving it. Sometimes the best ideas do not come from the top; they come from the floor, the front line, and the team you lead.

The more senior you become, the further removed you are from the hands-on reality of the organization. That is not a flaw; it is simply a fact. If you are not careful, it can lead to blind spots. You remain grounded in reality by listening to those closest to day-to-day operations. You remain grounded by speaking last in meetings, allowing you to hear the comments and suggestions of the team.

I have seen the contrast between listening and arrogance play out repeatedly in both the military and corporate worlds. In the Army, I was taught the chain of command and to follow orders without question. That mindset exists for a reason. In combat, one does not have time for debate. Orders must be executed because lives depend on immediate action. However, here is the truth that many do not discuss: that model breaks down outside of combat. In more than two decades of service, fewer than 10% of the orders I gave were in life-or-death scenarios. So why do so many leaders carry that mindset into every situation?

I believe it is often rooted in a fear of looking weak, a fear of being wrong, or a fear of losing control. However, the irony is this: when leaders choose to listen, ask for input, and explain the *why*, they do not lose control; they gain influence. They build credibility.

I once worked for a company where my boss did not listen to others. He believed he was the most intelligent person in every room, always right, never wrong. His arrogance was not subtle. It was apparent to me and to everyone else who worked for him.

At the time, I led the company's safety and security efforts. I brought years of experience conducting internal investigations involving fraud, theft, safety violations, and employee misconduct. I knew how to uncover the truth, not through force, but through process, patience, and experience.

I briefed him on our caseload during one of my boss's visits. I explained how my interview with one suspect had led to the identification of others, helping me piece together a larger picture. His first question was, "Did you get a confession?" I told him no, not yet. Without hesitation, he responded, "Then you are doing the interviews wrong."

I was stunned. It was not the questioning that bothered me; I welcomed challenges and different perspectives. What made me furious was his complete dismissal of my approach and of my experience. I tried to explain my process, the psychology behind interviewing, and how effective investigations are rarely about forcing confessions; they are about building evidence. He cut me off. He said, "Every time I conduct interviews, I get a confession."

That moment said everything. My boss was not interested in understanding. He was not curious. He had already decided that his way was the only way, even in a field without experience outside our company's environment. It was a powerful reminder that arrogance masquerading as confidence is one of the fastest ways a leader loses the respect of his or her team.

Great leaders do not need to be the smartest in the room; they must be smart enough to listen. Over time, morale dropped, initiative dried up, and people stopped speaking up because they knew it would not matter. I ultimately decided to leave the company.

Here is the irony many leaders miss: When you listen, ask for input, and explain the *why*, you do not lose control; you gain influence. You build credibility.

Commanders earned our respect not because they demanded it, but because they trusted us enough to listen. I have seen what happens when a leader asks, "What do you think?" A small question opens the door to trust, engagement, and better decisions. Listening does not make you less of a leader; it makes you more of one. In a war zone, where trust and speed can mean survival, listening becomes more than a leadership skill, it becomes necessary.

I worked for a leader in the Joint Operations Center in Afghanistan during sustained combat operations, a place where urgency was constant, and precision was everything. As a medical planner, I sat at the intersection of chaos and coordination, where information met action, and every decision carried weight. The room was humming with radio chatter and satellite feeds on a big-screen TV that would make any couch potato jealous. My job was to ensure that wounded soldiers had a chance, coordinating MEDEVAC routes, surgical capabilities, and evacuation priorities in real time. What made it possible, what made it *work*, was a leader who listened. He never dismissed concerns during operations and never assumed he had all the answers. He asked questions, trusted the expertise in the room, and created an environment where people like me could speak up and be heard. His leadership style maintained high morale and saved lives. In a place where seconds mattered and outcomes were unforgiving, listening became one of our most powerful tools.

During a complex operation, all senior leaders gathered in a side planning room. Tension was palpable. Some leaders had even traveled from Fort Bragg to join. The staff began the MDMP, the Military Decision-Making Process, a structured approach to analyze missions, compare courses of action, and coordinate execution.

As the planning process unfolded, each staff section took turns briefing the team, which was a regular part of the process. But then came a moment I will never forget. The team had nearly wrapped up our plan when the CG, the Commanding General, General McChrystal himself, turned to me and asked, "Do you have everything you need to be successful?"

I told him we did not. I laid out what was still missing, half-expecting some pushback. Instead, without hesitation, he said, "Then get it. And if you run into problems, let me or the CSM know."

I nodded, "Yes, sir," but inside I was stunned.

He did not just ask me. He went around the entire room, giving every staff leader the same opportunity. There were no rank barriers, no ego, just an elite commander, described by former Defense Secretary Robert Gates as "perhaps the finest warrior and leader in combat I ever met," genuinely seeking input from his team.

That moment reshaped my perspective on leadership. If my CG could ask for feedback, so could I, regardless of position or title. And I have done precisely that ever since.

Let me be clear: Leadership is not a democracy. You still have to make the final call. However, informed decisions are made by hearing all sides. The best leaders do not just give directions; they create conversations. Silence is not a weakness; it is a signal that you respect others enough to hear them. In that silence, authentic leadership begins.

Takeaway 1: Listening Builds Credibility and Influence

Leaders often believe that showing strength and assertiveness defines leadership, as this is what they observe. This story changes the narrative about listening and the fear many leaders have regarding what strengthens a leader's ability to lead. In reality, listening builds something much more sustainable: influence. By asking for input and explaining the rationale behind decisions, leaders do not lose control; they gain the trust and buy-in of their teams. That buy-in translates into stronger execution and alignment because the team feels heard and valued. When leaders show that every voice matters, they build a culture of shared purpose and understanding, becoming one of their greatest assets in any organization. Who on your team has insights that you might be missing? How can you create space to hear them this week?

Takeaway 2: Your Position Does Not Guarantee Respect, Your Behavior Does

The military has a rigid hierarchy; it is easy to assume that rank alone should command respect. This is no different in the corporate role; someone from the C-suite should command respect due to their position. But this story illustrates a more profound truth: Respect must be earned through consistent and respectful behavior. The contrast between my squad leader and General McChrystal could not be starker. While one used rank to silence others, the other used his position to open doors for dialogue. Genuine respect grows not from authority, but from authenticity. When leaders trust others and listen genuinely, they build credibility that no title can achieve alone. In what ways can you demonstrate authenticity rather than relying on your title?

Takeaway 3: Silence Can Be Strategic

When used well, silence is a sign of strength, not passivity. The silence that followed the CSM's revelation of my credentials spoke volumes. It was a moment that reset expectations without needing confrontation. Effective leaders understand when to speak and when to hold space. Strategic silence allows others to reflect, encourages thoughtful dialogue, and demonstrates humility. For example, leaders who speak last in meetings benefit from hearing diverse perspectives before sharing their own. Silence is not an absence; it is presence with purpose. What might your team say if they knew you were finally ready to listen?

Takeaway 4: Arrogance Is the Fastest Way to Lose a Team

My corporate experience is a sobering example of how quickly leadership can falter when arrogance takes the wheel. When leaders believe they have nothing to learn, they shut down opportunities for growth for themselves and their team. This can cause severe damage to the organization's culture. Dismissing experienced voices can damage morale, breed resentment, and foster groupthink. In contrast, humility keeps the door open for innovation and growth. The downfall of my boss was not his lack of intelligence, but his refusal to listen. That refusal cost him the trust and the energy of his team. Is your need to be right costing you the people you lead?

Takeaway 5: Leaders Create Conversations, Not Just Commands

The best leaders do not merely issue orders; they foster dialogue. Hear this and hear it clearly: "Leadership is not a democracy." It is essential to keep that distinction in mind. It means that while leaders must make the final call, those decisions are stronger when shaped by open communication. Creating conversations does not weaken authority; it strengthens outcomes. When people feel included in the process, they are more invested in the results. Leadership is as much about connection as it is about command, and those who master both inspire followership that lasts beyond any single mission or meeting.

Chapter Seven

Coach, Mentor, Leader

Knowing the Difference, Living All Three

L eadership is not a single role; it is a blend of voices, each showing up when needed most. Sometimes you need a coach to sharpen your skills, sometimes a mentor to expand your vision, and sometimes a leader to bear the weight of responsibility. Great teams thrive when these roles overlap, but clarity matters: knowing the difference between coaching, mentoring, and leading can be the difference between growth and stagnation, or success and failure.

Early in my career, my coach barked over the noise as he evaluated how we set up the aid station and drove us to perform the task exactly right. Hours later, my mentor walked with me, pointing out the subtle details I had missed, and the minor signs that often told the real story of what was to come. That night, my leader stood in the operations center, calm in the chaos, making decisions that would determine whether we all made it home.

Years later, I witnessed the same dynamic unfold in a Fortune 500 crisis meeting: the coach sharpening execution, the mentor shaping perspective, and the leader bearing the weight of the outcome. All three are vital, but they are not the same.

Yet sometimes they are. In the best cases, a single person may wear all three hats: driving performance like a coach, expanding vision like a mentor, and standing firm in responsibility like a leader. That overlap is rare, but when it occurs, it can change the trajectory of an individual, a team, or even an entire organization.

That difference is why understanding these roles matters. If you confuse these roles or fail to step into more than one when required, you risk missing what your people truly need from you.

Here is how I would break it down:

Coach

A coach does not need to be in your chain of command. Their focus is on performance within a specific skill or area. You do not work *for* a coach; you work *with* them to improve.

In the military, some of the best coaches were not in my chain of command at all. They were Non-Commissioned Officers correcting my form on the range or peers showing me how to use new equipment. Their authority did not come from rank but from expertise, and I am a better leader because they cared enough to coach me. Coaches demand repetition, discipline, and effort because they see not only who you are, but also who you could become.

Today, many organizations are rediscovering what the military has practiced for decades: coaching is not about hierarchy; it is about growth. In high-performing corporate environments, coaching is increasingly used to develop leaders, refine decision-making, and strengthen accountability, not as a corrective measure, but as an investment.

The most effective corporate coaches, like their military counterparts, do not rely on position or title. They rely on credibility, experience, and the willingness to tell the truth. They challenge assumptions, reinforce standards, and push people to operate at a level they may not yet believe they can reach.

The principle remains the same across uniforms and industries: people do not grow simply because someone outranks them. They grow because someone is willing to walk beside them, demand more, and refuse to let them accept less than their potential.

History provides us with this lesson. When Thomas Edison was a struggling telegraph operator, he met Franklin Leonard Pope, a seasoned inventor. Pope was not Edison's boss, yet he took him in, shared knowledge, and encouraged him to pursue invention seriously. He coached Edison not out of obligation, but out of belief in his potential, and that guidance helped shape Edison into one of history's most prolific inventors.

I observed this again when I worked in the corporate world. The Continuous Skills Development (CSD) leads were the coaches on the floor. They showed me and all the other new employees the basics, corrected mistakes, and drilled fundamentals until performance became second nature. What made them stand out was consistency. They did not just train new associates; they reinforced skills, identified gaps, and helped people grow every day. Over time, that built more than technical competence; it created a culture in which learning was normal, and development was expected.

Looking back, I realize the power of coaching. A coach shapes performance in the moment, but a *great* coach also lays the foundation for long-term growth, sometimes doubling as a mentor when teaching broadens perspective, and even stepping into leadership by setting the standard for the entire team.

Sharpening skills is only one part of the journey. Once you can perform, the bigger question becomes where you are headed. That is where mentors step in.

Mentor

If a coach works on your performance, a mentor works on your direction. A mentor does not just ask, "Are you doing it right?" They ask, "Are you doing the right thing?"

A mentor relationship is voluntary and personal. They care about your long-term growth, your career path, and sometimes your life outside work. They may not hold authority over you, but they invest because they wish to see you succeed. Mentors provide perspective by helping you see beyond your current situation.

Throughout my military career, I have had multiple mentors who guided me at different stages, but one of the most impactful came later in my career: Colonel (Ret.) Dwayne Wagner. He had held a variety of leadership positions and positioned himself as someone who genuinely invested in the development of officers across the Army.

I first connected with him through social media because we were in the same branch. I occasionally messaged him with questions, and he always responded with thoughtful advice. When I finally met him in person at the Command and General Staff College in Fort Leavenworth, where he served on the faculty, it was clear why so many officers considered him a mentor.

I never officially asked him to mentor me. Instead, we would meet for coffee, talk about our families, and in between, he would deliver honest, often difficult advice. He was not afraid to tell me things I did not want to hear, but I always knew that his words came from a place of genuine care.

At one point, I interviewed for a job while wrestling with the decision of whether to retire after completing that assignment. Dwayne advised me not to mention retirement during interviews. However, when I was asked about my five-year plan, I mentioned retirement anyway. I did not think much of it until I did not get the job.

When we talked afterward, he did not say, "I told you so." He simply reminded me of the conversation we had and then helped me refocus on new opportunities. With his encouragement, I adjusted my approach, interviewed again, and obtained the next position.

What makes Dwayne's mentorship so powerful is that it did not stop with me; he got to know my family and even guided my daughter. Over time, he became more than a mentor; he became part of the family. While coaches shape performance and mentors shape direction, leaders bear the ultimate weight. They do not just influence your growth; they hold responsibility for people, missions, and outcomes. That responsibility changes everything.

Mentors expand your vision beyond the immediate task at hand. They do not just help you with the problem in front of you; they help you see what is ahead, what is possible, and who you are becoming.

Leader

Then there is the leader, the one who steps up when it is no longer about practice or potential but about people and mission.

A leader is connected to position and responsibility. They set direction, allocate resources, and bear accountability for outcomes. You work for them in a formal sense, but the best leaders go beyond authority; they build trust and inspire you to give more than what is written in your job description.

I did not realize it until I started looking back, but I was one of the luckiest people alive. I did not have many poor leaders in my life. That made choosing only one to highlight in this section a difficult task. However, one name kept coming to mind: Sergeant Major Thomas Perez, call sign "Pineapple." What made Pineapple extraordinary was not just his title or achievements. It was the manner in which he lived out three inseparable qualities of leadership: authority, culture, and responsibility.

I first met Pineapple when he was the detachment sergeant for the Headquarters Company of the 44th Medical Brigade. Later, we worked together for several years at the Joint Special Operations Command at Fort Bragg. Over time, he advanced to the rank of Sergeant Major. He was inducted into the SOMA (Special Operations Medical Association) Hall of Fame, a testament to the impact he had on the special operations medical community.

What made Pineapple remarkable was not just his accomplishments. It was the manner in which he carried authority, shaped culture, and assumed responsibility.

Authority

Pineapple always said, "When in charge, take charge." That phrase sounds simple, but in practice, it is what distinguishes leaders who crumble under pressure from those who thrive. I learned that lesson during deployment while serving as part of the operations team. I was tasked with developing a plan, and the first version that I presented was not well received by everyone. Some of the senior personnel pushed back firmly, instructing me to develop a better proposal. It would have been easy to back down or second-guess myself, but I held firm. I continued to work on the issue, refining it where I could, but I did not flinch or walk away from the decision I had made. Several hours later, the word came down that we were proceeding with the plan that I had created. At first, I thought I had convinced them. Later, I realized it was not my people skills at all. It was Pineapple, working quietly behind the scenes, lending his authority to me when I did not even know it. He never said a word about it, but I knew.

That taught me that absolute authority does not have to shout; it speaks by empowering others.

Culture

Pineapple also understood the power of culture. Our team consisted of soldiers who had not been raised in the special operations community. That situation can be intimidating, with differing standards, different expectations, and a reputation for not always welcoming outsiders. However, Pineapple knew how to bridge that gap. He built a culture of trust that extended far beyond our small detachment. He ensured that we were not merely "attached" to special operations units; we were part of them. I will never forget my first major training exercise with one of the units that we supported. Their leadership did not know me, had not worked with me before, and had no reason to trust me, except for one factor: I worked for Pineapple. For them, that was sufficient. His reputation had already established the introduction for each one of us. He had built a culture to which we all strove to adhere, a standard that compelled us to give everything we had on every mission.

Responsibility

Then there was responsibility. At one point, Pineapple had to deploy, and he placed me in charge of the team in his absence. That responsibility involved representing our detachment at a senior leader event filled with command sergeants major, colonels, and several general officers. Nervous, I half-joked with him beforehand that I would simply grab the coffee and remain silent. Pineapple smiled but did not say much. Within fifteen minutes of the conference starting, a significant decision was on the table, and the senior command sergeant major turned directly to me and asked, "What do you think, Doc?" In that instant, it became clear what he had done. Pineapple had transferred his responsibility to me. He was not merely giving me a seat at the table; he was trusting me to contribute to the discussion. Once again, he was correct. I provided my assessment, the discussion progressed, and our team was adequately represented because Pineapple trusted me sufficiently to assume his responsibilities in his absence.

Pineapple led with a quiet confidence that few could emulate. He carried the **authority** to make difficult decisions, shaped the **culture** that enabled us to exceed our own expectations, and assumed **responsibility** that could never be deferred or ignored. He modeled all of these qualities without ever making it about himself. Leaders like him do not merely impact missions; they shape individuals. If you were fortunate enough to serve with Pineapple, you emerged as a better leader. In Pineapple, I observed all three roles: he coached us in skill, mentored us in vision, and led us with responsibility. Pineapple embodied what we all hope to find in leadership: someone capable of coaching, mentoring, and leading, occasionally all simultaneously.

Now here is the truth: not everyone begins with equal access to these roles. Some people are born into environments in which mentors and coaches are naturally integrated. Consider this: is it surprising that a Manning descendant is the starting quarterback for the Texas Longhorns? Is it surprising that Bronny James is also in the NBA, and his brother Bryce is playing at a D1 program?

That does not imply their path was easy or that they lack talent in their respective fields, but it does indicate that they had an advantage. For most of us, we have to go seek out the appropriate mentors and coaches. Success is never easy, irrespective of your name or family background, but having the right coaches and mentors significantly enhances the likelihood of advancing professionally.

This principle applies in any profession. If you wished to open a restaurant, you would not begin by referring to yourself as a Michelin-starred chef. You would start by working in a kitchen, learning from an experienced chef, and participating in exercises that frequently involved late nights and extended hours. Everyone aspires to reach the top, but you only attain that position by beginning at the bottom.

Here is something most people overlook: these relationships are not one-sided. A coach, mentor, or leader is making an investment in you. Ask yourself this: Do you deserve that investment? Are you worthy of mentorship? Why should someone choose to devote their time, energy, and wisdom to you?

The truth is that people with the power to transform another's life will not squander their efforts on someone unprepared to do the work.

Within the Army, Non-Commissioned Officers adhere to a creed: *"All soldiers are entitled to outstanding leadership, and I will provide that leadership."* As a leader, leadership is your responsibility. Coaching and mentoring are voluntary commitments. If you want people to be your coach or mentor, you must be worthy of such guidance. Ask yourself: Are you worthy of mentorship? Are you coming to work prepared to learn and contribute as a competent teammate? Do you own your mistakes and learn from them? If you practice these things, competent coaches and mentors will seek you. Conversely, if you do not have a mentor, ask yourself why.

The truth is that the best leaders play all three roles at different times.

1. When your people lack skill, they need a coach.

2. When they lack perspective, they need a mentor.

3. When the pressure is on and the stakes are real, they need a leader.

If you miss that distinction, you will either overwhelm them with advice, abandon them to figure things out alone, or hide when they need your presence most.

Mirror Test

So here is the mirror test:

1. Who is coaching you right now, sharpening your skills?

2. Who is mentoring you, helping you see the bigger picture?

3. Who is leading you, someone you would trust when the pressure hits?

Now flip it:

1. Whom are you currently coaching?

2. Whom are you currently mentoring?

3. Whom are you currently leading?

If you cannot answer these questions with names, you might be in trouble, so make filling in those names your mission. Growth is not just about how high you climb; it is about how many people you bring with you. Coaching sharpens skills. Mentoring shapes character. Leadership takes ownership.

At the end of the day, leadership is not about choosing to only be a coach, a mentor, or a leader. It is about knowing when to sharpen your people, when to provide perspective, and when to take the weight and say, "Follow me."

I think back to those days when the air was thick with diesel and dust. In that environment, we did not have the luxury of selecting one role; we had to be all three. A coach, to maintain and refine skills. A mentor, to expand the vision beyond the immediate engagement. A leader, to assume responsibility when the bullets were genuine, not simulated.

That same balance applies in corporate offices, boardrooms, and communities. If you only coach, your people may improve their skills but fail to see the bigger picture. If you only mentor, they may see possibilities but never execute. If you only lead, they may follow, but they will never grow.

The leaders who leave the deepest mark are the ones who can do all three and know when each voice is needed. That is the kind of leader worth following.

Chapter Eight

From Peer to Leader

The Hardest Promotion

Being promoted over your peers is one of the most challenging tasks a leader can face. The friendships and camaraderie you once shared now sit alongside new responsibilities, heightened accountability, and the need to make difficult decisions. Balancing empathy with authority is never easy, but it is essential.

When I first became a squad leader, responsible for about ten soldiers, I believed the best way to lead was to be liked. These were the same men I had trained with, laughed with, and built strong bonds alongside just weeks earlier. I did not want to break that connection, so I kept things light, avoided hard conversations, and let performance issues slide.

Then came our mass casualty (MASCAL) training. We were working in a field hospital, triaging incoming casualties as they arrived. One of my soldiers skipped a safety step, something we all knew not to overlook. Afterward, I pulled him aside.

His response? "Relax, man. You know I have got it. We are good." He was not being disrespectful; he was talking to me like a buddy. That was the problem. I had blurred the lines. I had been creating friendships instead of establishing leadership, an issue that became evident during that exercise.

Our performance was not a disaster, but it was not good either. Later, our senior enlisted advisor gave me a direct and honest critique. It stung, but he was right. That day, I learned something that has stuck with me ever since: In leadership, clarity must come before comfort.

You can care deeply about your people, but once your authority becomes negotiable, so do their discipline, safety, and trust. In any leadership role, especially when promoted from within your job, the role is not to be everyone's buddy; it is to lead them. That does not mean cutting people off; it means ensuring the relationship aligns with your responsibility to the team and the mission. The lines between friend and supervisor can blur quickly. However, regardless of the setting, military or corporate, those lines matter. It falls to the leader to make those lines crystal clear.

So how do you keep leadership and friendship from colliding?

Know Your Role

When you step into leadership, clarity is your first responsibility. You are no longer just a teammate; you are the guide, decision-maker, and point of accountability.

During MASCAL training, a soldier skipped a critical safety step. I hesitated, trying to stay "buddy-like," but it cost our team's performance. That day, I learned that clarity must come before comfort. Your role is to lead, not to be liked.

That does not mean cutting people off or losing connection. It means realigning the relationship to fit the weight of your responsibility to the team and the mission. You are not only responsible for your own output but also for setting direction, making decisions, and ensuring the entire team delivers. That shift requires clarity in what you do and how you show up each day.

So what does that look like? First, you must understand the mission. What is the big picture? Why does your team exist, and how does that align with the organization's objectives?

Main Effort, Supporting Effort

In military planning, everything revolves around the main effort, the primary mission that must succeed for the operation to succeed. Supporting efforts remain significant, but they exist to support the main effort. This concept translates effectively to the corporate world. Suppose your company plans a large-scale expansion into a new region. Logistics becomes the primary focus, as everything depends on product movement, supply chain alignment, and warehouse readiness. That is the center of gravity.

However, to make that possible, HR must serve as a supporting effort. Human resources recruit, onboard, and train staff to execute the logistics plan. Security must understand the movement schedule to protect the company's assets at both locations. They are not off-mission; rather, they are on a support mission to ensure the primary objective is completed.

The problem in many organizations is that everyone believes they are the main effort. That is when resources become scattered, priorities compete, and leaders lose alignment. Not only must the leader understand the main and supporting efforts, but the entire team must also do so.

Great leaders define the main effort clearly and empower supporting efforts to support it, rather than compete with it. However, this only works if leaders in supporting roles understand the entire organization, not merely their specific area of responsibility. Human resources, IT, finance, and other enabling functions must understand how their work fuels the mission, not only how to execute their tasks. Understanding the organization end to end allows them to anticipate needs, adapt quickly, and contribute meaningfully.

Why Does This Matter

When someone steps into leadership, especially after being promoted from within, it is easy to assume they can lead in the same manner they worked before, as "one of the team." But that is a trap. Leaders are no longer responsible solely for tasks; they are also accountable for people, direction, alignment, and results. If they fail to recognize this shift, they risk confusing the team, blurring lines of accountability, and stalling progress. Approach your new role with humility, clarity, and confidence. Begin by acknowledging the shift from day one. Let the team know that while relationships remain important, responsibilities have changed, and your role is now to lead them.

Set new expectations early and do it openly.

- "I have moved into this new role and will take it seriously."

- "You have worked with me as a peer; now I want to earn your trust as your leader."

- "I will always respect our connection, but I must hold the line where necessary."

Bottom line

Leadership is not about distancing yourself from the team; it is about defining your new role clearly. Leaders who do not understand their role cannot create clarity for others. When clarity is missing, trust erodes, silos form, and execution suffers. However, when leaders fully embrace their responsibilities, defining the main effort, aligning supporting efforts, and leading with the mission in mind, the entire organization moves in sync. That is when momentum builds, performance improves, and leadership earns absolute respect.

Know Your People

What are your team's strengths and weaknesses? Examine your team's strengths not only under ideal conditions but also when circumstances are less than optimal. As the leader, identify which tasks they enjoy most, and incorporate tools such as *StrengthsFinder*[1] or *DiSC* assessments into your hiring process[2]. Identify areas where they struggle, not to criticize them but to coach them into improvement. Watch for tasks that team members consistently avoid or where outcomes fall short of expectations. Understand how each team member prefers to work: some thrive under a clear structure, others with creative freedom. Here is a suggestion that most leaders do not follow. Ask simple questions such as, "Do you prefer detailed plans or big-picture goals?" or "What kind of feedback helps you most?" Finally, determine what motivates each team member. Is it recognition, challenge, autonomy, mission accomplishment, or purpose? Knowing what motivates them allows you to lead with greater empathy, precision, and trust. When your team knows you see them as individuals rather than as roles, they will give greater effort and be more honest.

---

1. Tom Rath, *StrengthsFinder 2.0* (New York: Gallup Press, 2007).

2. *DiSC® Assessment*, Everything DiSC, accessed November 12, 2025, https://www.everythingdisc.com/ .

Why Does This Matter

Knowing your people is not just a good idea; it is the foundation for trust, growth, and performance. When you understand what your team does well, what they need help with, and what motivates them, you can lead with precision rather than guesswork. You will delegate more effectively, coach more skillfully, and avoid costly misunderstandings.

More importantly, your people will feel seen, not just for what they do, but for who they are. That trust becomes a bridge to accountability. People are far more likely to give their best effort, take feedback seriously, and step up during challenging moments when they know that their leader understands and values them. This ongoing process takes time, and leaders must be patient while working toward this goal.

Provide Clarity with the Five Ws, and Confirm It with a Brief Back

Communication is a cornerstone of leadership. When giving instructions, use the **Who, What, When, Where, Why**, and always request a brief back to confirm understanding.

During an investigation, I asked a junior leader to conduct interviews. Instead of assuming he understood, I requested a brief back. This simple step prevented miscommunication and reinforced accountability.

The brief back is essential because it closes the loop on communication. As a leader, it is not enough to give guidance; you must confirm that it was understood as you intended. Too often, people tell the leader that they understand, but what they mean is that they know it from their own perspective, not the leader's. The brief back allows your team to repeat their understanding of the task, timeline, responsibilities, and desired outcomes.

Why Does This Matter

Eliminates Assumptions

Even the clearest-sounding instructions can be misinterpreted. A brief review exposes gaps or misunderstandings before they become errors.

Example

You say, "Make sure the report is ready by Friday," but in the brief back, the team says, "We will have the draft by Friday." That one word, "draft", shows that there is a disconnect. Without the brief, you would not catch it until it is too late, resulting in an angry leader and a confused worker.

Builds Accountability

When someone repeats the plan in their own words, they are not just showing that they heard you, they are stepping up, taking ownership, and declaring, "I got this." That is where real commitment begins.

Develop Confidence and Competence

Over time, brief back-and-forth interactions help teams learn how to internalize guidance, ask clarifying questions, and plan more effectively. It becomes part of a culture of high performance.

Saves Time and Frustration

It is easier to fix a miscommunication upfront than to redo work later. The brief back is a simple, fast way to prevent confusion from snowballing into lost time or failed outcomes.

Bottom Line

The brief back is not about micromanagement; it is about mutual clarity. It shows that you care enough to verify understanding as a leader, not just issue orders. That is how strong teams build trust, stay aligned, and deliver results.

Accountability

Once expectations have been clearly set, the next step is equally important: meeting them. This is a clear definition of accountability. This is where accountability becomes a test of your leadership. If executed correctly, you will gain your team's respect. If executed incorrectly, it can unravel the fabric of your organization. You must stand by your word if you tell someone they will be held accountable for their actions. This can be a pivotal moment in a leader's journey. It is not about micromanaging; it is about ensuring that performance standards are upheld and actions have consequences. If someone consistently misses deadlines or underperforms, you must address the issue rather than avoid it. If you ask your team who the best performer is, they will answer instantly. Ask who the weakest link is; most will not hesitate to respond. If you consistently let poor performance slide, your team will question whether you are in charge or merely trying to keep the peace. You will break your team's trust as a leader if you do not consistently hold underperformers accountable.

Making the Tough Calls Is Part of the Job.

As a former military officer, people often assume that "tough calls" involve life-or-death decisions in combat. However, that is not the whole story. Leadership requires tough calls in all kinds of situations, including personnel decisions, resource allocations, and policies that affect people's lives. My responsibility is to lead, not to be liked. However, that does not mean these decisions are easy.

When I served as a company commander at Guantanamo Bay, Cuba, we had a strong team and a demanding mission. Soldiers could bring their families for a two-year tour or go unaccompanied for one year. My wife and I chose the unaccompanied option so our children could remain in school.

One day, a soldier approached me under the open-door policy. She was a single mother whose leave request had been denied, and she wanted to go home for her son's birthday. She broke down in my office, pleading her case, and as a father, I felt every word. I wanted to say yes; however, I had to consider the big picture. Our mission required strict control over personnel on the island. I had to find the fairest and most consistent way to manage leave.

One of the best leaders I have ever worked with, Darcey Overbey, once told me, "Sometimes you have to be ruthless, but you must be equally ruthless across the board." She also said, "If these decisions ever become easy, it is time to ask yourself why you are still leading."

That moment reminded me that leadership involves difficult decisions, especially those concerning people you care about. When making these tough calls, consider these three areas to guide your decisions:

- **Policy**: What does your organization say regarding it?

- **Precedent**: What occurred the last time a similar situation arose?

- **Perception**: How will others view this decision?

These three questions are not a checklist: they are a compass. When emotions are high and the situation is complex, these guidelines will help you remain grounded in principle rather than succumb to pressure. They allow you to lead with fairness and consistency, even when the answer is not what someone wants to hear.

In that situation, I had to implement a policy in which once-in-a-lifetime events, such as weddings or graduations, took priority over annual events, such as birthdays or anniversaries. It was not perfect, but it was consistent. Over time, the team came to understand that decisions were based on fairness rather than favoritism.

Ultimately, people may not always agree with your decisions, but they will respect how you made them. Leadership is not about avoiding hard calls; it is about making them with clarity, fairness, and empathy. You do not have to be heartless to lead; you just have to be honest. If your team sees that your decisions come from principle rather than preference, they will trust you even when the answer is no.

Ask yourself: Have I ever let a relationship blur my leadership role? Have I hesitated to correct, redirect, or challenge someone because I did not want to lose their friendship?

The truth is that ships can sail side by side. However, when leadership and friendship collide, one always takes on water.

Choose leadership.

Chapter Nine

Leadership Discretion

The Spirit vs. the Letter of the Law

L eadership in the Gray

Leadership is not tested when everything goes according to plan; it is tested when the book does not have the answer.

During my military career, I learned that policies and procedures can take you only so far. The moments that truly defined leaders were not about enforcing rules; they were about navigating the gray, choosing between the letter of the law and the spirit of the law. That is where judgment, courage, and character come into play.

This chapter examines how leaders exercise discretion, not to circumvent rules but to uphold values, protect individuals, and foster a strong culture.

Let us be honest: anyone can follow a checklist; anyone can quote policy. But reality begins when the rules fall short.

During my military career, I have watched leaders rise and fall based not on their ability to enforce policies but on their ability to understand people. I have seen situations where the right decision was not in the manual; it was in the moment when a leader responded. Whether they defaulted to the letter of the law or leaned into its spirit made all the difference.

I have been in rooms where junior leaders asked for permission to do what they already knew was right but lacked the authority. I have been on the other end of a decision where policy said one thing, but my gut and leadership experience said another.

This chapter is about those moments, the moments that test our values, stretch our judgment, and define our legacy.

Leadership discretion is not a license to ignore the rules. It is a responsibility to lead with wisdom, courage, and empathy. It involves knowing when to enforce and when to elevate, when to say "no" and when to say, "This time, we will do what is right, not what is written."

Let us discuss how to lead in the gray area.

Real leadership does not always occur in black-and-white decisions. When you are in charge, you sometimes must make tough decisions in the gray area. During my time in the military, we referred to it as your left and right limits, the range in which you had the discretion to make one choice or another. We were stationed overseas when one of my soldiers received the terrible news that his wife, who was pregnant, had been rushed to the hospital because of pregnancy complications. At the time, we were outside the United States, so coordinating his return and reunion with his family was challenging, but we were able to make it happen. The baby was born prematurely and placed in the NICU, but remained stable. Due to the soldier's leave, known as paid time off (PTO), he returned to the unit after approximately one week. After returning to the unit for one day, the baby's condition declined rapidly, and everyone was concerned she would not survive. We reached the point at which the soldier was about to run out of leave days. We wanted to support the soldier, but without leave, he would have needed to return. I spoke with a friend who was at the same base where the baby was receiving care, and he suggested a temporary change of station. It was not against policy, but it was in the spirit of caring for someone in a terrible situation. We completed the paperwork and temporarily assigned him to a unit, allowing him to support his wife and ill child without using all his leave days. I wish I could say that she survived, but unfortunately, the baby passed away. Thanks to a team of leaders, of which I was proud to be a part, our soldier was able to spend that time with his family.

That experience showed me something I have carried ever since: leaders must know the difference between enforcing what is written and honoring why it was written. This brings us to the heart of the issue: the letter of the law versus the spirit of the law.

Understanding the Difference Between the Spirit and the Letter

Rules matter. They bring order, set standards, and provide a baseline of fairness. However, if leadership were as simple as enforcing rules word for word, any computer could do it. Real leadership, human leadership, requires discernment. This requires understanding the difference between the *letter* of the law and the *spirit* of the law.

The Letter of the Law

The letter of the law is the literal interpretation. It is black and white. It is what is written in the policy manual, the standard operating procedures, and the handbook. Let us be clear: it is not wrong to follow it. Structure provides people with confidence and clarity.

However, this is where it becomes dangerous: leaders may lean so heavily on the letter that they forget *why* the rule was originally written. "This is the policy" becomes the default answer even in moments when the policy does not serve the people it was designed to protect.

I have seen this firsthand in both military and corporate environments. A soldier misses formation by one minute and is reprimanded without question. However, no one asks *why* he was late. It turns out that he was helping a new patient get to the medical clinic for a serious issue. Technically, he had broken a rule. Morally, he showed initiative and compassion. Punishing him sends the wrong message.

The Spirit of the Law

The spirit of the law asks a different question: "What is the intent behind this rule?" It requires leaders to understand the *why* behind the *what*. This is where values come in: compassion, fairness, accountability, and wisdom. These do not always appear in the fine print, but they form the foundation of trust.

Leaders who operate with the spirit of the law in mind are not rule-breakers; they are principle-driven. They see rules as tools to guide behavior, not as cages to trap it. They are not looking for loopholes; they are looking for what is right.

When a leader bends a rule to do the right thing, people remember it. When they rigidly enforce a rule to protect themselves or assert control, people remember that too. It is one thing to understand the difference; it is another to recognize why it matters so deeply for culture, trust, and the mission.

Why It Matters

Every organization experiences moments when rules and reality collide. This is where organizational culture is either built or broken. If you want a culture of trust, initiative, and ownership, you cannot lead like a traffic cop. You must lead as a human being.

The goal is not to ignore rules; it is to *interpret them wisely*. That is discretion, and it is one of the most important tools a leader can have. Your people are always observing how you apply discretion. Are you fair? Are you consistent? Do you care more about policies or about people?

You can set the tone for a healthy organization in which values guide behavior, not merely checklists and compliance.

Leadership is rarely demonstrated in easy decisions. It is found in the gray areas, the ones that challenge your judgment and test your character. That is where the difference between the letter and the spirit becomes clear. And that is where your people find out what kind of leader you are.

How Military and Corporate Environments Often Differ, but Both Need Discretion

Letter vs. Spirit: A Lesson in Holiday Discretion

I knew that when I transitioned out of the military, my leadership style would need to adjust. When I first transitioned from the military to the corporate world, one of the earliest lessons I learned came not from a boardroom but from the break room, and it had everything to do with time off.

At the company where I worked, as in many organizations, seniority dictated who received approval for PTO requests, especially around the holidays. The policy was clear: If two employees requested the same time off, the more senior employee would be granted leave. On paper, the policy was clean, consistent, and enforceable. However, when I began to unpack it, I hit a wall. "If I am the newest person on the team," I asked, "and everyone wants Christmas off, does that mean I will never get Christmas as long as I am the new person?" Without hesitation, the answer was yes: that is the **letter of the law.**

It was a black-and-white policy in a world filled with gray areas. To me, it went against everything I understood about fairness, morale, and team cohesion. I was not asking for a loophole; I was asking for logic and common sense. In the military, we still followed policies, but there was always room for leaders to apply sound judgment, care for their people, and think long term about team cohesion.

The **spirit of the law**, in this case, would be to rotate the holiday schedule so that everyone, regardless of seniority, eventually has the opportunity to spend meaningful time with his or her family. Not only would that feel fair, but it would also build trust and demonstrate that leadership understood what mattered to employees beyond productivity metrics.

This experience reminded me that leadership discretion is not just about the rules; it is about the reasoning behind them. Policies are essential, but without compassion and common sense, they can become barriers to culture rather than its builders.

That lesson made me realize something important: there are times when you absolutely need the **letter of the law** and times when the **spirit must take the lead**. The skill is knowing which moment you are in.

When to Rely on the Letter, and When to Lean on the Spirit

There are moments when sticking to the letter of the law is necessary, especially in matters of compliance, safety, or legal protection. However, in many day-to-day decisions, leaders must ask a more human question: *What is the right thing to do for this team, in this context, while still upholding the organization's values?*

Discretion lives in the tension between those two poles. Great leaders do not just follow rules; they interpret them with empathy and fairness.

When Strict Rule-Following Becomes Leadership Avoidance

In one of my former organizations, morning physical training was a non-negotiable 6:30 a.m., rain or shine. That meant most people started their day around 5:00 a.m. or 5:30 a.m. to be on time for formation. It was part of our rhythm, our culture, and our discipline. However, leadership also requires the ability to flex and adapt to the reality of what your people are going through.

The Super Bowl was approaching, and we knew that most of our soldiers would be up late. Some would host gatherings, others would attend them, and nearly everyone would be watching the game. The Monday after the Super Bowl is practically a national holiday in spirit, if not in law.

I had a friend in a different unit who told me their leadership made a slight adjustment: instead of 6:30 a.m., they would do it in the afternoon, with a work call at 9:00 a.m. The same amount of work. The same standards. But it acknowledged reality, and it gave their people just a little grace.

We proposed the same idea to our platoon sergeant. His answer? "No. That is not how we do it." He did not raise it up the chain; he did not even ask. He followed the letter of the law, not the spirit of leadership.

Weeks later, during an all-hands meeting, someone else brought up the very same idea. The commander did not hesitate. "That sounds like a great idea," he said. "As long as we are getting physical training in, I am good with it."

It was not about the push-ups. It was about the principle. Our platoon sergeant was not protecting discipline; he was avoiding leadership. He was not following the rules; he was hiding behind them.

Do not get me wrong: rules matter. Consistency matters. But when a leader refuses to even ask the question, when they default to "that is the way we have always done it," what they are saying is, "I am afraid to lead." That moment revealed the danger of hiding behind rules instead of leading through them. It also set up one of the most essential truths about discretion.

Why Discretion Matters

Discretion is not something good leaders merely talk about; it is an actionable skill, and it is a leadership responsibility. Without it, leaders tend to default into one of two dangerous extremes: rigidity or recklessness. Both can cause significant harm to an organization and its people.

Let us start with **rigidity**. That is when leaders cling so tightly to the rules, policies, or standard procedures that they lose sight of the people those rules are meant to serve and protect. It is the manager who refuses to let a junior team member swap holiday leave, even once, because "that is not how we do things here." It is the supervisor who sees a clear opportunity to improve a process but will not raise the issue because it is not their place. It is leadership by fear of stepping out of line.

On the other side of the spectrum lies **recklessness**. That is when leaders ignore guidelines, push past boundaries, and make decisions based purely on urgency or emotion. There is no accountability, no consistency, and no consideration of the long-term consequences. One minute, they are breaking protocol to meet a deadline. Next, they are issuing exceptions so frequently that the team stops taking direction seriously.

In both cases, trust is lost. One kind of leader becomes resented for being inflexible. The other becomes resented for being unpredictable. Neither builds a high-performing team.

Discretion sits between those two extremes. Good leaders do not become experts at juggling them overnight; it will take time. It is not about playing favorites or making emotional calls. It is about using judgment. It is about asking: Does applying this rule in this moment serve the mission and the people? Does bending the rule still uphold our values, or does it damage them?

Discretion is about honoring the **spirit of the law**, not just the letter. And when it is applied well, it builds something rare: trust. People feel seen. They feel led. And when people think that, they will run through walls for you.

But if they are afraid that every decision will either be a brick wall of red tape or a wild guess from the hip, they will stop investing. They will stop owning their roles. And worst of all, they will stop believing in the leader. The best leaders do not stumble into this balance by accident; they build it through honest reflection and deliberate practice. Start by asking yourself: Have you ever followed a rule that did more harm than good? Where in my leadership style do I tend to lean, toward rigidity or recklessness? Do your people feel empowered to bring forward their ideas, or are they afraid they will be shut down?

Rules matter, but so does flexibility. Leadership begins where the rulebook ends. That is where your character, courage, and judgment come into play. That is where your people need you, not merely as a policy enforcer, but as a trusted, human leader.

When the Right Call Is Not the Popular One

A few years ago, I was responsible for logistics at my company. As part of that responsibility, I was required to conduct scheduled inventories of our assigned equipment and report the results to our higher headquarters. During one such inventory, I was notified that a key piece of equipment had gone missing.

For context: this was not something minor. It was a critical item, and if it were genuinely gone, it would reflect poorly on our company, especially on my commander.

My instinct was immediate: we needed to act. I wanted to bring the inventory team back in and begin a focused search immediately. However, a prescheduled event was scheduled for the next morning, and the battalion commander's standing guidance was that everyone was to leave early in preparation. When I informed my commander, her response was simple: "We will worry about it tomorrow."

Her answer stunned me. I knew this was not a problem that would age well overnight. More than that, I knew this would fall on her shoulders if it did not get resolved. I sat with her decision for a moment and then made my own: I called in the inventory team and ran a second, full accountability check.

It was not easy. People were not happy to come back in. But during that second check, we found the equipment. Crisis averted.

I believed I made the right call operationally, ethically, and even out of loyalty to my commander. But when I briefed her on what I had done, the reaction was not what I expected. The following week, I received a formal reprimand.

What did I learn from that moment? That discretion has a cost. But it also has a purpose. Sometimes, leadership is not about following the rulebook to the letter; it is about protecting your people, regardless of their actions or wishes, your mission, and your integrity, even when it means standing alone. If discretion costs something at the individual level, it builds something at the organizational level: culture. And few companies illustrate that better than Nordstrom.

Discretion in Action: How Executives Empower Their Teams

Leadership is not just about knowing when to apply the letter of the law or when to lead with its spirit; it is about building a culture where others are trusted to do the same. We have discussed senior leaders exercising discretion, but something transformational occurs when everyday associates feel empowered to use sound judgment without fear of punishment. That level of trust does not just reflect policy; it reflects culture. And when the culture supports discretion at every level, you unlock a kind of leadership that does not need a title to act. One company that has consistently modeled this is Nordstrom.

In corporate leadership circles, one of the most celebrated anecdotes about discretionary leadership centers on **Nordstrom. W**hile most know it as the *tire-return story*, what they may not realize is how it frames the CEO's approach to culture and leadership.

Nordstrom's famous story, in which a customer returned snow tires to a Fairbanks, Alaska, store even though Nordstrom does not sell tires, is more than just a quirky tale. It serves as a visible marker of how the organization's **top leadership** guided decision-making throughout the company.

Former CEO **Bruce Nordstrom** and his successor generations made it clear, from the boardroom down, that empowering employees **to use judgment** was not only allowed but expected.

Employee manuals were famously slim: a single 5 × 7 card stating:

"Use your best judgment in all situations. There will be no additional rules."

Under that philosophy, store associates operated with discretion, even if it meant overriding the expected policy. When Craig Trounce issued the tire refund, it was not an ad hoc decision; it was backed by a cultural expectation set from the top: **culture over compliance**. Even though it was not standardized, the CEO and the leadership did not punish it; they reinforced it.

Leaders such as Bruce Nordstrom and, later, co-presidents Pete and Erik, emphasized that **culture shapes behavior**, not just policies. Bruce described his legacy as "waiting on the customer," a principle that shaped every leader's discretion. All decisions, exceptional or everyday, were rooted in a shared set of values: humility, trust, respect, and initiative.

This approach communicates a clear message:

"Do right by the customer even when the rulebook does not say so."

Rear-end anecdotes or unusual refund requests were not just tolerated; they were elevated as teaching moments. Customers told the story. Employees shared it. It became legendary not because of its rarity but because leadership reinforced it. What Nordstrom shows us is that discretion rooted in values creates freedom without chaos. Here is how leaders can put that into practice.

Building a Culture of Disciplined Discretion

Discretion without discipline is chaos. But discipline without discretion is stagnation. The key to effective leadership is not choosing one or the other; it is building a culture that equips leaders at every level to exercise *disciplined discretion*: the ability to make the right call in the moment, aligned with values rather than merely rules.

Train for Judgment, Not Just Compliance

Most organizations train leaders to follow procedures, but few train them to think critically when procedures do not cover the situation. If we want leaders who can make tough decisions under pressure, we must *train them to make sound judgments*. This does not mean abandoning standards; it means giving leaders the tools to interpret them responsibly.

Suggestions for Organizations to Consider

- Use scenario-based training that mirrors real-world gray areas, not only black-and-white checklists.

- Encourage after-action reviews (AARs) that focus on decision-making, not only outcomes. A good AAR can be difficult, so if you leave one feeling great and believing you made all the right decisions, it wasn't a good AAR.

- Reinforce values first. When leaders understand *why* a policy exists, they can better determine *when* it applies. That is why, early in this book, we discuss purpose and the underlying reason behind the work;

it is not merely a catchphrase but the foundation of a successful organization.

Reward Thoughtful Judgment

If you want a culture of discretion, you must *recognize and reward it*. People repeat what is celebrated. When a junior leader makes a difficult decision that goes against the grain but aligns with your organization's mission, acknowledge it. Make it clear that thoughtful initiative matters.

However, be cautious: rewarding *results* without evaluating the underlying reasoning can backfire. Reckless decisions that succeed due to luck are not wins. Build systems that celebrate critical thinking, especially when decisions are difficult or unpopular.

Examples

- Include "judgment" in your evaluation metrics, not only performance.

- Create spaces (such as leader development forums) where discretion stories are shared and learned.

- Show publicly that you support your leaders when they make a reasonable call, even if it does not lead to perfect results.

Empower Without Eroding Standards

The fear some leaders have about discretion is absolute: "If I give too much freedom, I will lose control." However, when discretion is tied to clear expectations and reinforced by mentorship, it strengthens standards.

Empowerment does not mean an open door to do whatever feels right; it means trusting people to act within a known framework. Discipline is still required. The difference is, it is *internally* driven by commitment, not just compliance. Set left and right limits, define what is non-negotiable and where flexibility exists. Teach leaders how to ask the right questions before taking action: Is this aligned with our purpose? Does it support the team? Will it maintain trust? Do not just correct. When a leader misses the mark, use it as a teaching moment, not only a disciplinary one.

Disciplined discretion is not soft leadership. It is some of the most challenging leadership there is. However, when executed well, it creates resilient and agile teams that are ready to navigate complexity, and leaders who not only follow rules but also carry the weight of responsibility with **wisdom**.

Leadership discretion is not about breaking rules; it is about applying them with **wisdom**. Lean too far on the letter of the law, and you risk rigidity that crushes initiative. Lean too far on the spirit of the law without boundaries, and you invite recklessness. The balance point is disciplined discretion: knowing when to enforce and when to adapt, and always anchoring decisions in organizational values.

At the end of the day, your people will not only remember the policies you enforced. They will remember the **judgment** you demonstrated. They will remember whether you cared enough to see them, not just the rulebook. That is the type of leadership that builds trust, shapes culture, and leaves a lasting **legacy**.

Chapter Ten

Excellence by Design

What Great Organizations Do Correctly

W hy do some organizations soar while others fail? Have you ever wondered why some organizations are successful while others are not? In every great organization I have observed, whether in the military or the private sector, consistent qualities set them apart. Organizations do not become exceptional by chance. Excellence always begins in the same place: who is brought through the door. Recruitment is not merely about filling positions; it is the foundation of everything that follows.

Recruitment: Excellence Begins at the Entrance

Over my thirty-plus-year career in the U.S. Army, I served eight years in the Special Operations community. That experience became a benchmark for me; an example of what is considered "right" looks like when building a world-class team. This observation does not diminish the accomplishments of regular Army units; however, having served in both provides me with a unique perspective.

Understanding Special Operations

If you have never served in the military, allow me to paint a clear picture of what it means to be part of the Special Operations community. These are not your typical military teams. They represent the most elite, highly trained, and selectively chosen individuals in the United States Armed Forces: men and women entrusted with the most complex, high-risk, and strategically essential missions to defend the nation. When the President of the United States (PO-TUS) dials 911, the call is received at the headquarters of the Special Operations Command.

Special Operations units are not limited to but do include the Army Rangers, Green Berets, Navy SEALs, MARSOC, Air Force Pararescue, and the 160th Special Operations Aviation Regiment. They are called in when the stakes are highest, the environment is most unpredictable, and the margin for error is nonexistent. These professionals are trained to fight, think, lead, adapt, and solve real-world problems under extreme pressure.

What sets these units apart is not just their advanced gear or tough physical training standards, although both are formidable. Unlike most military assignments, in which soldiers are assigned to a specific location under permanent change-of-station (PCS) orders, entry into Special Operations requires a rigorous selection process. You do not get "assigned" to these units; you earn your place. That distinction makes all the difference. However, what truly makes Special Operations different is not just the missions or the gear, but how people gain entry. That is where selection makes all the difference.

Why Selection Matters

Selection into Special Operations is intentionally challenging because these roles demand more than mere competence; they require character. Individuals in these units are expected to perform independently while remaining deeply connected to their teams. Ego has no place in Special Operations. What matters most is performance, trust, and execution.

The lessons I learned in the Special Operations community shaped how I see leadership, accountability, and team dynamics. It is where I first understood what a high-trust, high-performance culture truly looks like and why replicating that mindset in the corporate world is both possible, powerful, and essential for success. That culture of selection and trust is not unique to the military. Some of the world's best companies share this principle. Netflix is a prime example.

Imagine a company where every employee is hand-selected, not simply based on their résumé; they are also committed to excellence, take initiative, and are held to the highest standard, not just because leadership demands it, but because their teammates expect nothing less when they join the team.

1. Tough Standards Protect Culture Before They Protect Performance

Rigorous selection is not about elitism; it is about preservation. When companies lower standards to fill seats quickly, they may solve a short-term staffing problem, but they create a long-term cultural one. High-performing cultures are fragile. They are built on trust, shared expectations, and the belief that everyone has earned their place on the team.

Special Operations units understand this instinctively. Selection is designed to ensure that those who make it through can be trusted when no one is watching. The same principle applies in business. When employees know their peers were selected through demanding, values-based standards, trust forms faster, accountability rises, and mediocrity has nowhere to hide.

2. High Standards Reduce Management Friction and Increase Execution

When selection is done well, leadership becomes simpler, not easier, but clearer. Leaders spend less time correcting basic behaviors and more time focused on strategy, development, and execution. Tough standards filter for people who take ownership, respond to feedback, and operate with discipline.

In Special Operations, leaders are not micromanaging fundamentals in the field; those expectations were set and enforced during selection. Companies that adopt similarly realistic standards benefit in the same way. Fewer performance issues, fewer interpersonal conflicts, and faster decision-making are not accidents; they are the byproducts of selecting people who are already wired for responsibility.

3. Tough Selection Signals What the Organization Truly Values

What an organization tolerates during hiring is what it will live with every day after. Selection is leadership's first and loudest message about standards. When companies are clear and uncompromising about what it takes to earn a seat on the team, they attract people who want to be held accountable and repel those looking for the path of least resistance.

This is why companies like Netflix emphasize talent density and high expectations. They understand that success is not built by policies alone, but by people who believe the standard is worth meeting. In both military units and corporate teams, excellence begins long before the first task is assigned. It begins with who is allowed through the door.

Would You Fight to Keep Them?

That is the question Netflix asks its managers about every team member in what they call the "Keeper Test." It is a standard, not a suggestion, and it drives one of the highest-performing cultures in business. If the answer is no, the employee is given a generous severance and released from the company.

Netflix does not rely on rigid rules or performance improvement plans. Instead, it expects employees to act like owners, not renters. They trust adults to be responsible and expect nothing short of excellence.

Much like Special Operations teams, their success is not an accident; it is the result of selective hiring, ruthless clarity, and relentless accountability. People do not stay at Netflix because they are merely exemplary. They stay because they are exceptional.

People, both veterans and business leaders, often ask me what it was like to serve in Special Operations. My answer is always the same. It was not about the extra pay or advanced equipment. It was not the high-profile missions, although those were certainly intense. The real difference, the defining factor, was the people. These were the best men and women the Department of Defense had to offer. What does this mean for leaders outside Silicon Valley or the military? The lesson is simple: selection creates standards.

What Business Can Learn

Why is this important for organizations outside the military? Because selection creates standards. When you make it more challenging to join your team, you do not simply filter out the unqualified; you attract a different caliber of professional. These are individuals who **do not want to be average**. They want to be elite. They seek challenge, accountability, and growth. It does not matter if your company sells construction equipment or installs security alarms in corporate buildings; the quality of your team is what matters.

There is nothing wrong with expecting more from those you hire. Building a team that consistently delivers at a high level should be required. If you set the bar high, you will attract those **committed to excellence,** not merely those seeking a paycheck.

Leadership: Preparation Over Promotion

Leadership in the Special Operations community is no accident. It is **intentional, practiced, and proven.** If you want to lead in one of these elite units, you do not get the job solely because of potential; you earn it because you have demonstrated performance elsewhere. For example, in the 75th Ranger Regiment, you do not enter as a new platoon leader fresh from the academy or ROTC. You must have already successfully led a platoon in a conventional unit. The same applies to company and battalion command.

This approach may sound demanding, but it is built on a clear truth: leadership in high-stakes environments requires preparation, not improvisation. There is no room for "learning as you go" when lives and missions are at stake.

That same truth holds in the business world: exceptional performance in one role does not guarantee success in another, especially when the new role involves leading others. Time and again, companies promote top individual contributors into leadership positions, assuming their past success will carry over, but often it does not.

Consider sports as a parallel. Only 20–30% of former professional athletes who become coaches are deemed successful. The remainder, roughly 70–80%, struggle, and many are dismissed within a few seasons because of poor records. Being great at the game does not automatically mean you can lead others through it. The skill set that wins on the field does not always build winning teams.

Leadership is a skill set, not a reward. Without intentional development, even the most talented employees can struggle in management.

Steps Every Organization Can Take

1. **Develop your own leaders.** Establish structured, intentional training programs that equip people to lead before they are promoted. Make completion of that training a requirement before assuming leadership

roles. It must be an organizational priority; otherwise, something else will always take precedence over training.

2. **Hire leaders with proven leadership experience.** If you bring someone in to manage a team, ensure they have led a team successfully or have completed a credible leadership-development program.

High-performing organizations do not leave leadership to chance. They prepare, mentor, and support leaders **before** those leaders are given responsibility. Even with the best people, culture does not sustain itself. Leadership is what turns high standards into lasting environments that people *want to join*.

Creating an Organization People Want to Join

At its core, a great organization is one that people want to be part of, not just for compensation but also for its **culture, mission, and sense of purpose**. When your recruitment process reflects your values, leaders are trained and trusted, and excellence is the standard rather than the exception, you create an environment in which people do not merely show up; they commit. They take ownership. They raise the bar. Most importantly, *they stay*.

Ultimately, people do not follow paychecks; they follow a **purpose**. Build that into the bones of your organization, and you will not just attract talent; *you will build loyalty*.

Great organizations do not stumble into excellence; excellence is intentional. They set high standards at the door, prepare leaders before promoting them, and create a culture in which people *choose commitment* over compliance.

From the Special Operations community to companies such as Netflix, the pattern is clear: when leaders raise the bar, the right people rise to meet it. Recruitment determines who is admitted, **leadership shapes how these people grow, and culture decides whether they stay**.

Ultimately, people do not follow rules, titles, or paychecks; they follow trust, vision, and purpose. If you build that into the DNA of your organization, you will not just create a team that performs; *you will make one that endures*.

Chapter Eleven

Legacy or Limitation?

Y ears ago, I cannot pinpoint the specific date, but I heard a story that stayed with me not because it was profound but because it was painfully familiar within the business world.

A young woman was preparing dinner when she cut the ends off a roast before placing it in the pan. Her husband asked, "Why do you do that?"

She shrugged. "That is how my mother always did it."

Curious, the young woman asked her mother, who said, "That is how your grandmother did it." So the young woman called her grandmother, who laughed and said, "I did that only because my pan was too small." This is not just a quirky story; it is how "we have always done it" can appear in organizations large and small, from kitchens to the Army.

What began as a practical choice became an unquestioned routine. This is precisely how outdated business practices survive in the corporate world. I have seen leaders defend broken systems with the same sentence: "That is how we have always done it." The same answer always resurfaces when these practices are challenged: "That is how we have always done it."

A good example of this occurred when I was in the Army. There was an Army policy that outlined grooming standards, including haircuts, hair length, and the wearing of mustaches. The policy applies to everyone, and nowhere in it does the policy state that officers are permitted one standard while NCOs are permitted another. Somewhere along the way, an unwritten rule emerged that officers do not wear mustaches.

When the time came for me to attend OCS, everyone in my unit knew that I would be leaving. One of the officers congratulated me and asked when I would shave my mustache. I had heard about this old tradition but had never paid much attention to it because I was an NCO. However, when I looked around, I saw that no officer in my unit had a mustache. When I asked a few of them why, I was told that it was simply tradition. Is there a difference between a tradition and the phrase "we have always done it this way"? That is where leaders must draw a line: Is this a tradition worth preserving, or a practice we have outgrown?

When Tradition Becomes a Trap

Tradition and policy, although they are similar, can confuse and hinder progress within an organization. Tradition is preserved intentionally. It honors identity, builds cohesion, and reminds individuals of who they are at their best. The salute in the military is a proud tradition. When upheld properly, tradition serves the mission. However, broken policies or outdated practices are often followed by default, simply because "that is the way we have always done it." That is not discipline; that is complacency in uniform. In elite teams, the best leaders ask why before they follow. They protect meaningful traditions but refuse to defend processes that no longer serve their people or their purpose. Leadership means knowing when to honor the past and when to challenge it. The true mark of a leader is not how well they preserve the past; it is how boldly they prepare their people for what is next.

What Kodak and Blockbuster Teach Us About Change

In the early 1970s, an engineer at Kodak developed the first digital camera. It was revolutionary, capturing images electronically instead of on film. He proudly presented the invention to his leadership. Kodak was a billion-dollar giant when it chose to ignore the digital camera invented by its engineers because it threatened its film business, which was its primary source of revenue. The future was in their hands, and they buried it. Digital cameras have revolutionized photography, and Kodak could have cornered the market, but the company chose not to embrace new technology. Instead, it followed tradition and missed the revolution. Another similar situation in which a company failed because it remained committed to past practices is when it continued to repeat them.

Blockbuster did not fail because of Netflix; it failed because it refused to adapt. Blockbuster dominated the home entertainment industry at its peak, boasting thousands of stores and generating billions in revenue. I remember spending many Friday and Saturday nights searching for the perfect movie at a Blockbuster store. However, success made the company complacent. When Netflix introduced digital streaming in January 2007, Blockbuster doubled down on late fees and brick-and-mortar strategies rather than innovating with customer needs in mind. Netflix offered convenience, personalization, and a forward-thinking model, but Blockbuster dismissed it as a passing trend. Blockbuster even had the opportunity to purchase Netflix for $50 million and walked away. The truth is that Blockbuster did not lose because the market changed; the company failed because it did not change. In business, arrogance and nostalgia are a dangerous combination. Blockbuster is a case study of what happens when leaders cling to what worked yesterday and ignore what is required to win tomorrow.

Tradition does not outrank truth in high-performance organizations. The leaders who thrive are not the ones who protect what has always been done. They are bold enough to ask, "Is this still right?" Authentic leadership is not about guarding the past. It is about preparing for what is next. Both cases prove the same point: nostalgia is not a strategy. Leaders who cling to yesterday's success often forfeit tomorrow's relevance.

Break the Cycle: Avoiding the "We Have Always Done It This Way" Trap

As previously mentioned, one of the most dangerous phrases in any board meeting is "This is how we have always done it." It sounds harmless, but it is often where innovation goes to die and where relevance erodes. Tradition does not outrank truth in high-performance teams or elite businesses. Leaders who succeed do not just protect legacy systems; they question those systems. Here are some examples of how to avoid the trap.

Encourage Constructive Dissent

Leaders create a culture in which team members are encouraged to question the status quo without it being viewed as an attempt to usurp the leader's authority. Team members should be encouraged and expected to feel comfortable speaking up without fear of retribution. Teams should feel safe when challenging existing processes, but they must also offer viable alternatives. Otherwise, they are merely complaining and are not attempting to improve the team. The timing of alternative courses of action is also essential to workplace dynamics. This will differ by organization. Determine what works best for your team.

AAR: A Tool for Organizational Evolution

One of the most effective leadership tools is the After Action Review (AAR), a practice that was ingrained in me during my time in the Army. It did not matter whether it was a tactical combat mission or planning for a military ball; the premise was the same: what was successful and what fell short of expectations. We gathered the team, stripped away rank and ego, and asked the hard questions, not to place blame but to grow as an organization.

Great organizations do not wait for a crisis to reflect; they create a structured process for reflection. The AAR is not merely a military tool; it is a technique that all organizations, regardless of size, can use to evaluate their operations. It is something every high-performance organization should adopt.

- **Restate the Objective**

   What were we trying to achieve? This question sets the stage. It ensures that everyone understands the original intent and the mission.

- **Review What Happened**

   Walk through the **facts**. Sequence the events. Capture what occurred, not opinions, but observable actions and measurable outcomes. This provides a common operating picture for all team members to observe. This approach works well, particularly when some team members are in different locations supporting the mission.

- **What Was Supposed to Happen**

   This is where leaders and teams take ownership. What did we do well? What did not go well? Both questions are essential. We celebrate what worked, but we do not shy away from what did not.

- **Identify Root Causes**

  Go deeper. Why did it occur in this manner? (Was it training? Communication? Planning? Resources? Culture? Leadership?) The goal is to uncover the truth, not excuses.

- **Extract Lessons Learned**

  Translate insight into wisdom. Identify specific takeaways that the team can apply to the next operation or project.

- **Define Actionable Changes**

  This is where improvement becomes real. What will we do differently moving forward? Are there policies, processes, or behaviors that require modification? If our leaders were not prepared, what steps would we take? What steps will we take if we do not have the proper equipment? If the team does not develop feasible changes, it will be in the same position next time.

During my Army days, the AAR was sacred because it reinforced a simple truth: learning is non-negotiable. Without continuous learning, progress stalls, and with stalled progress comes missed opportunities for growth and relevance. Success is not assumed; it is examined, refined, and repeated with intention. Failure, in that context, is simply information. The real risk emerges only when reflection is avoided and change is resisted. Leaders who create deliberate space for honest review cultivate teams that adapt, improve, and evolve faster than the environment around them. And in today's world, that pace matters.

Business leaders who ignore this tool miss a powerful opportunity. Reflection is not a luxury; it is a competitive edge.

Rotate Leadership Perspectives

Bring in different leaders from other areas. If an organization is large, bring in leaders from other locations to evaluate its operations. If an organization is smaller, bring in leaders from different departments. A marketing leader may observe what an operations team overlooks. This cross-functional insight exposes blind spots and fosters creative discussion and solutions.

Incentivize Innovation, Not Just Output

Most organizations reward only output. While results matter, they should not be the only achievements recognized. Leaders must remember that people want to know that their efforts matter. It does not always have to be a bonus or a significant award. Sometimes, it is simply calling someone out by name for their idea at a team meeting or recognizing a small achievement that reflects the correct mindset.

If an organization desires innovation, it must reward the courage to try something different. Google's renowned "20% time" allowed employees to explore new ideas. Not all ideas succeeded, but those that did became Gmail, Google Maps, and AdSense. That did not happen by accident. It occurred because the company rewarded initiative, not merely results.

Leadership sets the tone. When only safe choices are rewarded, risk-taking naturally disappears. But when effort, courage, and forward-thinking are recognized, teams begin to elevate the standard together. Innovation thrives in cultures where the attempt is valued alongside the outcome, and where growth is encouraged even before success arrives.

Invest in External Exposure

Encourage leaders to attend industry conferences, site visits, or mastermind groups. Observing how other organizations solve similar challenges often reveals where systems are stagnant or outdated. Do not allow these trips to become vacations at the organization's expense; require leaders to submit a report detailing what was learned and what should be applied to the team.

Legacy is not an excuse for stagnation or the avoidance of change. Great leaders know when tradition is a strength and when it is a crutch. If a practice exists in an organization only because it has always been there, it is probably time to examine it more closely.

For more information regarding dealing with Differences of Organizing Opinions (DOOO), see https://www.lisalinard.com/blog/dealing-with-dooo -differences-of-organizing-opinions

Chapter Twelve

The Moment Leadership Is Tested

T here are moments in a leader's life when the title becomes real, not only in authority, but also in burden. That moment did not arrive with the formal commissioning ceremony but in the quiet realizations and chaotic circumstances that tested everything I believed about responsibility, integrity, and courage. Becoming an officer was not about stepping into command but about unlearning my thought process as an NCO, embracing new challenges, and learning to lead when others hesitated. I did not know it then, but those early years would shape my military career and leadership philosophy for the future.

My first assignment as an officer was a defining moment. Even though I had served in the Army for eighteen years by this time, I was filled with apprehension, hoping that this change would shape my approach to leadership for the years to come. As an officer, my role in the Army shifted, requiring a distinct style of guidance and management. I had often heard cautionary tales from mentors of prior-service officers faltering, misled by the illusion that what earned them stripes as successful NCOs would seamlessly translate into officer success. Yet I knew better. I embraced the role of platoon leader, resisting the temptation to replicate the familiar patterns of a platoon sergeant. I was placed in the capable hands of an NCO, Joseph McAuliffe, affectionately known as "Mac," and I discovered a sense of balance. Despite his questionable taste in football teams, Mac, hailing from the Northeast, proved to be a trustworthy confidant, always offering candid recommendations. Our relationship lasted for years. Notably, Mac was not a Military Police like me; rather, he was a seasoned infantryman temporarily assigned to our unit due to a shortage of senior NCOs. Having previously served with his battalion commander, I advocated for Mac to join us when the need for a platoon sergeant arose. Even though I would not join the platoon for a while, I was still glad to know that I would eventually work with Mac.

Being assigned as a Military Police lieutenant in a non-MP unit, I felt anticipation regarding my inevitable transition to MP platoon leader. Before donning that mantle, I endured the obligatory tenure on battalion staff. Although staff work is vital for a unit's success, it paled compared to the crucible of leadership I sought. Amidst this, a unique project arrived: integrating working dogs into our unit, preparing us for impending deployment to support the Global War on Terror. This project was not a stroke of luck; it reached me through deliberate channels. The brigade commander, well-versed in the context of my background, personally endorsed me at the recommendation of his former battalion commander, T2, a legend from the Ranger community. This set a high bar for success, making it clear that failure was not an option, not only for the new dog program I would be in charge of but also for my career transition as an officer.

The deployment loomed, casting a shadow over my current responsibilities. Yet duty called, and I supported the brigade mission alongside our dog handlers. The soldiers from the MP platoon were consistently welcoming; some even asked when I would take over as platoon leader, and my answer was always the same. "When the Battalion Commander (BC) says I can." Even though we all knew I would arrive eventually, I was still teased about my stint as a "FOBIT," meaning I was confined to the safety of the base. Despite the relative security, the longing to lead patrols pulsed within me. I never encountered a soldier who wore the title of FOBIT as a badge of honor.

Our unit deployed, and I spent the initial part of the deployment certifying the dog teams at Bagram Air Base. Once that was complete, the dog teams deployed with various elements of our brigade task force, and I worked with the battalion staff to monitor progress. I also worked closely with another exceptional NCO, Patrick Fowler, a capable young NCO from the "Roll Tide State." Our task was to train the Afghan police to establish and operate an Emergency Operations Center (EOC). Staff Sergeant (SSG) Fowler helped design the training program, oversaw the development of the stadium seating in the EOC, and collaborated with the Navy Seabees during construction.

Finally, the call came to assume what I believed to be my rightful place as MP platoon leader. I was so excited that my enthusiasm could not be contained, yet I feared being unable to lead in this environment. I stepped into this pivotal role, determined to validate the investment in my officer training and the recommendations of Generals McChrystal and Thomas. Conducting a meticulous equipment inventory was a task bearing the weight of potential dismissal if not handled appropriately. Leaving the FOB to lead a genuine combat patrol symbolized liberation from the confines of safety.

I caught wind of rumors circulating within the platoon, suggesting discontent with the current platoon leader. However, I approached these murmurs with caution, recognizing that disapproval of leadership was not uncommon. How many times had you said that you disliked your superior's decisions? I understood that I might face similar scrutiny when my time to lead arrived. The unwritten rule was never to speak ill of another officer, and I sought to abide by it.

Accompanying the platoon on patrols was part of my left-seat, right-seat ride (orientation), so I could learn the responsibilities of a platoon leader from the incumbent. I absorbed every detail, carefully mapping the routes etched by countless missions. During one such patrol, a moment of reckoning arrived. Some members of the team's leadership were inside the Afghan police headquarters, discussing my arrival and future plans. Suddenly, a shot pierced the air, propelling us into action. Acting swiftly, I, one of my NCOs, and the other incumbent platoon leader exited the building to assess the situation.

Stepping out first, I encountered Sergeant Deitz, a Southern California native and one of the most capable leaders in the platoon. Inquiring about the source of the gunfire, he gestured toward a corner. Advancing cautiously, we encountered a soldier lying on the ground about sixty feet from our location, and we were initially unsure of his condition. Upon closer inspection, it became apparent that the soldier had experienced a negligent discharge from a .50 caliber weapon, meaning he fired a round unexpectedly. The incident resulted in a momentary state of shock, and he fell from the ASV. While calming the soldiers, I noted the absence of the incumbent Platoon Leader (PL), whom I had last thought was behind me. Although I refrained from speculating on the reasons for his absence, it was evident to the other soldiers that the PL was not present to assess the situation.

Following proper protocols, leadership was informed, and we were directed to return to our battalion base at FOB Shank, anticipating a serious debriefing. The negligent discharge was a significant matter. I was not personally concerned about a potential career-ending mistake, since I was not officially in charge.

Back at the base, conversations with my superiors hinted at underlying tensions and a lack of confidence in the incumbent PL. Although I harbored reservations about his leadership style, loyalty dictated that I support him.

During our second patrol, we met with a senior Afghan official, facilitating introductions with the other Afghan police officers with whom I would collaborate over the next year. After a daylong meeting, we returned to the base for the night. As we departed the Afghan compound, a sense of disorientation overcame me. Although I recognized the feeling, I understood that I was not in a leadership position and refrained from taking command, remaining silent on the radio. Unbeknownst to me, my driver, nicknamed Pop Tart (the origin of the nickname remains a mystery), informed me that another NCO in a different vehicle wished to speak with me. Using a separate channel known only to soldiers, they expressed concern about the situation and urged me to take command before any potential danger escalated. Assessing the circumstances and finding no immediate threat, I radioed the PL to assist, but he assured me that he had everything under control. Despite the prolonged twenty-minute journey, which extended to nearly two hours, I dismissed it as a standard navigation error. I reiterated to the NCO that I was not in a position of authority and encouraged him to support the PL, emphasizing that any recommendations could be addressed once I assumed command and assuring him that his concerns would be heard.

I maintained a good relationship with my company leadership and had the opportunity to speak with my company commander, who inquired about the status of operations. Initially, I offered reassurances, stating that operations were proceeding without issue. However, upon his insistence, I reiterated that considerable tension existed between the team and the incumbent. I was familiar with the person I would replace, yet I harbored reservations about his leadership style.

During our conversation, my commander hinted that he had heard similar concerns, and because I respected him, it was vital for me to be candid. I informed him that I believed the incumbent exhibited cowardice by not responding alongside me when we heard the gunshots and suspected potential danger. This admission prompted a lengthy discussion during which my commander emphasized the importance of reporting safety concerns and assured me of his willingness to take action if warranted.

Later, I learned that the PL, in response to an internal investigation, shifted blame onto me for the entire episode, despite being in a position of authority during the incident.

A few days later, we resumed our patrols, spending approximately two nights alongside our Afghan partners before returning to FOB Shank. By this point, the PL had exhausted all opportunities for redemption. As a second lieutenant, one is expected to make mistakes, yet also to learn from and avoid repetition. Leaders can instruct on tactics and planning, yet no leader can teach one common sense or how to act correctly, particularly when you think no one is watching. While on patrol, vigilance required that we remain locked and loaded, prepared for potential enemy engagements. Upon returning to the FOB, standard safety protocols required clearing all crew-served weapons and individual firearms.

During this routine procedure, I observed that the PL failed to disembark all soldiers from his MRAP and neglected to clear his crew-served weapon properly. Unfortunately for him, this oversight did not escape the notice of the Battalion Executive Officer (XO) and the Battalion Command Sergeant Major (CSM), who were conducting inspections at the gate. While guiding my vehicle to the clearing barrels, I was initially met with a harsh tone until the XO recognized that it was I. The XO was furious and instructed me to inform the incumbent to report to the XO's office once the vehicles were in position. I do not know what was said in the office, but I was informed that my transition was complete, the platoon was now under my charge, and I needed to complete the inventories as soon as possible.

Despite my initial assumptions about his proficiency as a PL, it became apparent that the platoon's culture demanded urgent reform. During my brief time with them, I observed that combat operations had fallen into a "state of routine," leading me to believe corrective action was necessary. My first leadership challenge was obtaining buy-in for my new perspective, as I had not led any patrols prior to this point. Nonetheless, this paradigm shift was significant, even though the team did not embrace this idea.

Upon conveying my concerns to senior leadership, unanimous agreement emerged that a return to basics was warranted before venturing from the FOB again. Despite initial pushback from the platoon, we spent several days revisiting fundamental procedures. Although I encountered from all ranks, including my NCOs, who questioned my authority after only a week of patrol experience, it became clear that entrenched habits required correction.

Patrol briefings, frequent practice of enemy contact drills, and nighttime exercises became routine as we diligently addressed shortcomings. Each soldier reviewed the Rules of Engagement (ROE) and, over time, became well-versed in them, executing react-to-contact drills with increasing proficiency. Gradually, I observed the repetition taking hold, transforming them into the cohesive unit I had envisioned. Satisfied with our progress, we set out on our inaugural patrol.

As we embarked on our inaugural patrol, with our destination being the Afghan police complex in Chark for a rendezvous with our Afghan counterparts, I wondered whether the team had fully embraced the new operational approach within the platoon. While I informed them of my intention to ride in the lead vehicle for navigation, I also aimed to demonstrate my commitment to leading from the front. I led a four-vehicle convoy, with the second vehicle hauling a trailer for a range session with our Afghan counterparts. Then chaos erupted not far from FOB Shank, and a loud explosion rocked our convoy. We executed our crew drills, which had been rehearsed recently. However, it was not the turmoil that caught my attention; rather, the disciplined response of the team impressed me most. We suffered a casualty and vehicle damage. Despite the failure to apprehend the perpetrator of the IED, their adept reaction affirmed their readiness. A subtle nod from SSG Will conveyed all the assurance I needed, indicating that the rigorous additional training had paid off.

Takeaway 1: Leaders Need to Be at the Point of Friction

Leaders cannot make assumptions from a distance. A leader must place themselves in a position where they can see clearly, evaluate the facts, and make the best decision. This is the responsibility of a leader: to step toward uncertainty, gather the truth, and provide direction when others await clarity.

The point of friction is where tension, resistance, and uncertainty live. It is where plans meet reality and where leadership is either proven or exposed. Leaders who remain removed from this space rely on secondhand information, filtered perspectives, and assumptions that often fail under pressure. Those who step into it gain context, credibility, and trust.

Leaders who avoid the point of friction quickly lose their team's confidence. Those who showed up on the range, in the field, or during difficult moments earned it. Presence signaled commitment. It told the team that decisions were being made with firsthand understanding, not convenience.

This principle translates directly to the corporate world. Whether the friction exists on a shop floor, in a tense meeting, or during organizational change, leaders who are present can identify problems early, remove obstacles, and make informed decisions. More importantly, their presence reassures teams that they are not alone when the work gets hard.

Leadership does not mean eliminating friction. It means standing where it exists long enough to understand it, own it, and guide others through it.

Takeaway 2: Buy-In Creates Shared Ownership

In high-performing organizations, buy-in is not a luxury; it is part of the foundation of an organization's operating structure, ensuring it is integrated into daily operations. When leaders secure genuine buy-in, they create an environment where team members do not just complete tasks; they take ownership of outcomes. This psychological shift transforms passive compliance into active engagement. Employees who feel heard and involved are more likely to take initiative, innovate independently, and hold themselves accountable. In this way, buy-in serves as the bridge between delegation and true empowerment. Leaders delegate tasks but never delegate responsibility. Corporate leaders must understand that shared ownership is not about relinquishing authority but about multiplying impact. Individuals who see their fingerprints on the solution protect, improve, and advocate for it. Leaders who cultivate this sense of co-authorship are those who build sustainable, resilient, and agile teams.

Takeaway 3: Buy-In Builds Trust Before It Is Needed

Trust is the currency of leadership, and, like any currency, it must be earned before it can be spent. Buy-in is one of the most powerful investments a leader can make in building trust. When leaders take the time to communicate transparently, involve others in decisions, and genuinely consider diverse viewpoints, they lay the groundwork for future cohesion. In moments of uncertainty or crisis, individuals do not follow a title; they follow someone they trust. Leaders who consistently pursue buy-in demonstrate emotional intelligence and foresight, thereby fostering firm loyalty under pressure. In high-stakes corporate environments, where rapid change is constant and organizations are dispersed, the leaders who thrive have preloaded trust through meaningful engagement. Trust and influence will be discussed in more detail later in the book.

Takeaway 4: Leadership Without Ego Leads to Better Decision Making

Yet the best leaders know when to set aside their ego in favor of wisdom. Leadership is not about being in charge, but about those in one's charge. Authentic leadership seeks outcomes, progress, and service for others. When decisions are made from a place of ego, they often serve personal ambition rather than the team's success. By contrast, when ego is checked at the door, leaders are more receptive to feedback, more collaborative in execution, and more likely to surround themselves with people who challenge them rather than simply echo their statements. For corporate executives, this humility creates an environment where ideas, rather than titles, carry the most weight, allowing innovation to emerge from all levels of the organization. A team needs to know what its members say, regardless of their position in the hierarchy. Ego may win attention in the short term, but humility sustains one's credibility over the long term. Yet buy-in and trust can crumble quickly if leaders allow ego to obstruct progress. I observed this play out firsthand with my predecessor, and it was equally destructive in corporate settings.

Takeaway 5: Great Leaders Know When to Allow Others to Lead

The IED explosion occurred behind me, which meant I was not in the best position to make immediate decisions. In that moment, I had to step back, let the situation unfold, and give my junior leaders the time and space to assess and take action. That is leadership maturity: sharing the spotlight and the responsibility.

Leaders driven by ego often hoard decision-making authority, fearing that their power will be diminished when shared. However, wise leaders recognize that empowering others at the right moment is both a developmental opportunity and a strategic advantage. It strengthens the team and extends the leader's influence far beyond any single decision.

Timing and trust intersect here: understanding when to pass the torch and trusting others to carry it forward. In that formation, I trusted every soldier to make the call if the responsibility fell to them. In high-performing organizations, military or corporate, this is not a weakness. It is how one builds a culture that will outlast their tenure.

The most respected leaders are not the ones who always have to step in. They are the ones who prepare others to succeed without them.

Leadership is rarely about being the most experienced or the loudest speaker in the room. It is about trust, consistency, and the willingness to act decisively when others are frozen by uncertainty. Ultimately, it was not my title that earned the platoon's respect; it was the repetition, the humility to listen, and the courage to lead from the front. I learned that some of the most significant leadership failures do not come from incompetence but from a lack of self-awareness and accountability. It lies in cultivating a culture where people feel safe to speak and to believe again in the purpose behind their mission.

Chapter Thirteen

Creating Change in an Organization

Have you ever worked somewhere and thought, *If I were in charge, this place would run much better?* Most people have. It is easy to imagine tightening standards, fixing inefficiencies, or eliminating poor practices when the responsibility rests on someone else. Observing from the sidelines is simple when the pressure of decisions and the consequences of failure rest on someone else.

Now imagine stepping into the role of the decision-maker, whether over an entire organization or a small team. The responsibility for improvement lands directly on the leader's shoulders. There is no more, *if I were in charge*, leadership is already in motion. What happens first? How does meaningful change take hold in a place accustomed to the status quo?

Change is more than authority; it requires mindset, strategy, and consistency. Many people resist change because comfort feels safer than growth. The challenge lies not only in identifying what must shift but also in guiding people through that shift in a way they can ultimately trust and support. Breaking the process into manageable steps can help.

Define the Span of Control

Before taking action, it is essential to establish the boundaries of responsibility. Where will the change occur? What area, team, or process falls within this leadership lane? In the military, this is known as a span of control, the area a leader is accountable for and can influence. In corporate terms, it is a team, a department, or an immediate sphere of authority. A leader may not run the entire company, but ownership over one corner of it is still significant.

No one has to transform the whole building at once. Change begins in the room where leadership already stands.

I had the opportunity to lead a small team of medics, and, honestly, it was one of the highlights of my career. For the first time, I had a say in who I worked with. We did not just select people based on rank or résumé; we ran them through a selection process. However, those tests were not just about how well they could perform medical tasks. They were an opportunity for me to get to know them as individuals. That process of putting them under pressure, watching them lead and follow, and seeing who they were brought us closer as a team.

We operated in a unique setup, rotating between two different commands every ninety days. Most teams do not deal with that kind of constant shift in leadership, expectations, and standards. However, we became used to it. And we did not just adapt; we looked for ways to influence the culture around us.

In one of the commands, we saw significant gaps. Morale, training standards, and discipline all needed a substantial boost. So we, as a team, decided to take what made our team strong, our unity, our mindset, and our shared goals, and attempt to spark change. We ran extra training, increased our physical training standards, and shared our energy with anyone willing to match it. And we encountered resistance. Every new idea seemed to meet immediate resistance.

But here is the thing about creating change: you do not stop when it becomes difficult. You keep going because change does not always appear in significant moments. It appears in small shifts. A few weeks in, we noticed people joining us for training sessions. Others began asking questions about our training. Standards slowly began to rise. The ripple effect was genuine.

Change does not always start at the top; it often begins with a group of committed individuals who believe things can improve. That was our team. And if you are trying to change your organization, do not wait for permission. Lead by example. Be consistent. And observe how your influence within your team spreads. *It is not about waiting for approval; it is about demonstrating what is possible.*

Is It a Problem or Merely Your Preference?

One of the most challenging questions a leader must ask himself or herself is this: *Is this a real problem, or is it merely my preference?* Yes, as a leader, you have the authority to change everything from the organizational logo to the way staff meetings are conducted. Early in my career, I thought leadership meant fixing everything that did not feel right to me. However, I learned that not every discomfort signals dysfunction; sometimes, it is merely a clash with my preferred way of operating. Real leadership means pausing to evaluate the issue objectively. When you, as a leader, are trying to differentiate between a problem and a preference, here are a few simple steps. First, ask yourself: Is this affecting mission success? If the outcome remains strong, it may not require adjustment, even if you do not like the process.

Second, how does the team feel about the process? Is it causing frustration, burnout, or disengagement among team members?

Third, is it harming your team's performance? Are they required to work harder and expend valuable energy and resources to operate in this manner? Is it time to implement a change or continue pushing forward?

Lastly, gather feedback from the team that will be involved in the process to understand their perspectives. Sometimes, the people closest to the work can help you see what is needed. If the system still functions effectively and the team is thriving, it might not require change.

I once attempted to overhaul a reporting process merely because it did not align with my style. Thankfully, I listened, adjusted, and learned a valuable lesson. Effective leadership means stepping back, seeking input, and aligning your decisions with the team's needs rather than your personal preferences. Leadership, in the end, is not about getting your way; it is about ensuring that you get it right.

At the heart of leadership is the ability to lead *yourself* before leading others. That means checking your ego, examining your motives, and asking the hard questions, such as whether the change you are promoting is truly necessary or merely more comfortable for you. The moment you begin making decisions based on your comfort, you will lose your team, but if you base change on what is best for the mission and the team, you transition from managing to leading. Great leaders do not merely drive change; they discern *which* changes are worth driving. That discernment earns trust, builds influence, and creates lasting impact.

Involve the Right People Early

If you want your ideas to take root and endure, you must involve the people who will live with the results. It is not enough to consult them; you need to invite them into the process as co-builders, not merely end users.

You may have sought their opinions in step two, but step three requires their **action**. This is where buy-in becomes ownership. People are far more likely to support what they have helped build because the outcome now reflects their fingerprints, not merely your vision.

This step is also essential for sustaining the change. The truth is that leadership changes frequently. People rotate out. However, when you involve the right people early, those with influence, institutional knowledge, and frontline experience, they become the keepers of the change when you are no longer present to champion it. That is how change becomes culture.

Do not be afraid to bring a few skeptics to the table. It may feel uncomfortable, but sometimes your biggest doubters become your most honest source of feedback, and, later, your most unexpected allies. When people feel heard and valued, even if they disagree at first, they are more likely to support the mission.

Here is a key principle: Do not present your change as a finished product. Present it as a shared solution. When people feel as though they are part of the story, they will help you write the ending. When a change feels as though it belongs to the team, they will work to protect it long after you have moved on.

After World War II, Europe was devastated economically, politically, and socially. Many leaders at the time debated how to help rebuild it, but few had a clear plan. That is when General George C. Marshall, then United States Secretary of State, stepped forward with a bold idea: to provide economic aid to help rebuild European nations and prevent future conflict. Marshall did not simply dictate a plan from Washington. He understood something many leaders miss: *lasting change requires shared ownership.*

Before the Marshall Plan was officially implemented, Marshall and his team involved European leaders from multiple nations. He did not present the plan as an American solution to a European problem. Instead, he framed it as a **partnership**, asking, "What do *you* need to rebuild your countries?" By involving the right people early, the individuals who would live with the results, he not only built support but also created long-term investment from each nation involved.

The result was over $13 billion in aid distributed across sixteen countries, a stabilized European economy, and one of the most successful foreign aid programs in history. More importantly, the Marshall Plan helped prevent the spread of communism and strengthened alliances that still exist today.

What made it work was not only the funding, but also the collaboration. Marshall did not force change; he invited participation. He knew that no matter how good a plan is, if people are not part of it, they will not carry it forward.

That is leadership at scale, and the principle is the same whether one is rebuilding a continent or revamping an organization. If one wants ideas to work and endure, one needs the people who will live with the change to help build it. Shared input creates shared ownership, and shared ownership sustains change.

Model the Change Both Publicly and Privately Before You Mandate It

You cannot expect people to adopt what you do not support. Too often, when a policy change comes down from corporate, some leaders will say, "We are only doing this because corporate says so." When people speak with you in the break room about the change, you mumble that you think it is a poor idea. How do you think that affects the team's perceptions? Does that sound as though you have confidence in the plan? Does it seem that you are supporting the change? Absolutely not. If you are not helping to bring about change through your words and actions, it will not happen.

You, as the leader, are the embodiment of the new normal and the change in the plan. Want more discipline? Be punctual. Want better documentation?

Complete your documentation first. Want a positive culture? Cease gossiping and begin recognizing good work.

In 2008, Starbucks was in trouble. Sales were declining, the customer experience was slipping, and the brand had begun to feel more like fast food than the premium coffeehouse experience that Howard Schultz had envisioned. Howard Schultz, Starbucks' former CEO, returned to lead the company during the crisis. One of his first moves was both bold and symbolic: he closed every Starbucks store in the country for three hours to retrain baristas on how to prepare quality espresso.

Did Schultz need to be in every store with a training manual? No. He publicly supported and privately championed the change. He did not say, "We are only doing this because the board says so." He took ownership. He visited stores, spoke directly with employees, and reaffirmed the reasons behind the reset. Behind the scenes, he invested in leadership development, prioritized employee engagement, and made culture, not coffee, the focus.

Schultz did not merely issue a memo from headquarters; he embodied the new direction. He knew that leaders are the walking, talking embodiment of the new normal. If he did not believe in the cultural reset, no one else would. When your words state, "We support this," but your actions in the break room suggest otherwise, the team hears the truth. Real change endures only when leaders model it both publicly and privately.

Expect Resistance but Remain Consistent

To be frank, saying the "F-word" in a workplace will almost always spark a reaction no different than mentioning change. Mention "change" in an organization and watch the energy shift: body language tightens, eyes roll, sighs surface, and resistance emerges. Change threatens routines. It challenges comfort zones. It disrupts the familiar patterns that people rely on.

Here is the truth: when there is no resistance, change is not being led. It is comfort being maintained. Real change creates disruption, and it is meant to. This is the moment when leadership earns its weight. This is where a flag gets planted, footing is secured, and forward movement begins.

Pushback should never be mistaken for failure. It rarely signals a wrong move; more often, it indicates work that matters. When the process becomes difficult, four anchors can help steady the course:

- **Remain consistent.** Change is difficult for all who are undergoing it.

- **Reinforce the "why."** People can endure a difficult "what" if they understand the "why."

- **Celebrate early wins.** Momentum builds when people observe success.

- **Adjust if necessary, but do not abandon the mission.** Flexibility is prudent; surrender is not.

Change is a fight worth undertaking. Do not expect a standing ovation at the outset. Lead regardless.

Change tests every dimension of leadership. It stretches patience, challenges communication, and places pressure on consistency. It can prompt second-guessing whether the pace is too fast or too slow or whether any action should have been taken in the first place. Those doubts are normal. But remember this: change rarely fails because the idea is flawed; it fails when the leadership behind it is weak.

Lasting transformation cannot be achieved by rolling out a policy and hoping for the best. It must be led with intention, visibility, and repetition. That includes understanding the span of control, distinguishing genuine issues from personal preferences, involving key stakeholders early, modeling the behavior before requiring it, and anticipating resistance, not as a red flag, but as a natural stage in the process.

Leadership during change is not about popularity; it is about trust. Teams watch closely to see whether words and actions align. They pay attention to how the change is discussed in the breakroom, not only how it is presented in the boardroom. If followership is expected, the example must come first.

This moment is more than an operational shift; it is an opportunity to strengthen a culture that supports growth. When resistance appears, and it will, stand firm. Do not retreat. Lean in, lead with resolve, and remember: the goal is not to preserve the status quo but to advance the organization.

Chapter Fourteen

Creating Value in Your Organization

I still remember the first time I truly felt seen by a leader. But just as clearly, I remember what it felt like to be nothing more than a name on a roster. My first assignment after coming home from Desert Storm was to a ground ambulance company at Fort Benning. I was not alone; several of my battle buddies from the war ended up there, too, just by luck. We were fresh privates, not carrying much except for the standard-issue duffel bags and the weight of experience most people did not expect from soldiers our age.

When we arrived, we headed to the orderly room for in-processing. Because the office was small, we were sent in groups. My last name began with "A", so I was the first one up.

As I was filling out paperwork, a sergeant walked in and said, with no attempt to hide her annoyance, "Where did all these privates come from?" The clerk replied, "Most of them just came back from Desert Storm. A few would be assigned to your platoon."

The sergeant rolled her eyes and sighed. "I already have enough work. I do not want any more privates than I need."

Then, from behind me, a calm voice spoke up: "I will take them all if you do not want them."

That voice belonged to Staff Sergeant Geraldine Ronan. She stepped forward, looked us in the eye, and said, "Welcome." It was just one word, but it changed everything.

The two sergeants stepped into the back room for a few minutes. When they returned, Sergeant Ronan pointed at me and said, "Grab your bags. You are with me."

Within twenty-four hours, I knew I had drawn the lucky straw. She did not merely take me under her wing; she taught me what it looked like to make people feel valued.

That single moment of being chosen, of being welcomed, has stayed with me for *decades*. Value is not about pay or position; it is about how you make someone feel from the very beginning.

After serving thirty-three years in the Army, I have come to believe that one truth about leadership, above all else, is that people stay in organizations where they feel valued. It is not where they are paid the most or where the perks are the best; it is where they are seen, heard, and respected.

Maya Angelou captured this perfectly when she said, "People will forget what you said, people will forget what you did, but people will never forget how you made them feel." That is not merely a nice quote to put on the wall; it is a leadership philosophy.

In every great team that I have been part of, whether in Afghanistan, Germany, or back home, there was one common thread: people felt like they mattered. They did not just have a seat at the table; they had a voice. And when people believe their voice matters, they lean in. They work harder. They care more. They become invested not just in their job but also in the mission.

So, how do we, as leaders, create that kind of value in our organizations? I will break it down.

Value Their Voice, Even When You Disagree

While in command of a law enforcement detachment at Fort Gordon in Augusta, Georgia, I had the privilege of working for an old friend from my Special Operations days, Sam Anderson. Sam was what we call a *PT stud*. He did not just pass physical training; he trained for triathlons in his free time.

Fitness was not just a part of his life; it was his lifestyle. When he organized a base-wide triathlon and later qualified to compete in the Ironman World Championship in Kona, Hawaii, it was clear: *what mattered to him needed to matter to us*. That meant I needed to improve our unit's physical training standards.

We created an incentive policy. Anyone who scored 270 or above (out of a 300-point PT test) would earn a reward. It was simple. However, not everyone saw it that way.

One of my new sergeants came to my office and voiced her concern. "Sir," she said, "this policy discourages those who are just struggling to pass. What if the incentive were based on how much someone improved, not only on how high they scored?"

I initially disagreed with her suggestion. In my mind, the standard was the standard. However, I also believed in listening. I talked it over with my First Sergeant, and together, we decided to implement her idea alongside ours.

We rolled out the revised policy and ensured that everyone was aware of its origin. It was not about the policy; it was about the message: *your voice matters here, even when it challenges the boss.*

That is how you create value by listening, adapting, and leading with respect.

Let me be honest: this part makes some seasoned leaders uncomfortable. After all, many of us came up in organizations where rank was everything, and "Because I said so" was a common phrase.

But here is the reality: You do not have to agree with someone to show them respect. If someone on your team brings up an alternative course of action, do not dismiss it out of hand. Do not default to, "That is not how we do things." Avoid the four worst words in leadership: "Because I said so."

Instead, respond with curiosity. Ask questions. If you ultimately decide to go in another direction, follow up and explain why. You do not need to justify every decision, but you do need to close the loop on it. That is how you show people that their input was genuinely considered.

One of my soldiers once introduced me to a completely different approach to solving a logistics issue. I did not end up using his plan, but I sat down with him and explained why. A year later, he told me that the moment was a turning point in his career, not because his idea had not been implemented, but because someone had finally taken him seriously.

Speak Last, You Might Learn Something

Another simple but powerful leadership habit is to let your team speak before you do. When you are in a room full of smart, capable people, do not start by sharing your plan and then asking for feedback. That puts pressure on people to agree, even if they have a better idea.

Instead, pose the problem and then ask your team: "What do you think?" Let everyone speak. Listen actively. Then, once they have shared their thoughts, share your plan.

Great leaders know they might not be the most intelligent person in the room, and they are okay with that.

Respect Their Time as Much as Your Own

Time is one of the most undervalued currencies in leadership. When you consistently arrive late to meetings, frequently reschedule one-on-ones, or waste people's time with unclear directions, you are telling your team that your time matters more than theirs.

I once worked under a boss who had a habit of showing up 5–10 minutes late to every staff meeting. At first, people waited patiently. Then, they started trickling in late as well. One day, he showed up on time, and half the team was still not there. He was furious. "Why are they not here?" he asked.

I did not hold back: "Because you have taught them that starting on time does not matter." To his credit, he listened. He changed. And the tone of those meetings changed with him.

Leadership is not defined by your words alone; it is defined by the example you set.

Recognize Contributions (Loudly and Often)

One of the simplest ways to make people feel valued? Tell them, publicly and privately.

Too many leaders fall into the trap of thinking, *They know they are doing a good job.* Maybe. But recognition is not just about acknowledgment; it is about motivation. It reinforces the behaviors you want to see more of.

I had a habit when I was a commander: anytime a soldier went above and beyond, I would write a personal note to them. Not an email. A handwritten note. You would be amazed at how much that small act meant. Years later, I would run into soldiers who still had those notes.

I had a brilliant NCO on my team who was responsible for ensuring we remained compliant, not just with Army regulations but also with federal policies. When it came time for our biannual inspection, thanks to his preparation and diligence, we achieved a perfect rating, a milestone our organization had never previously accomplished.

I made sure everyone knew who deserved the credit. Publicly and privately, I thanked him and let the team know that, without his expertise, we would never have reached that milestone.

Know What Matters to Them

Creating value means understanding what your people care about. That starts by asking: What are their goals? What are they struggling with? What motivates them?

You do not have to solve all their problems. But when you take the time to learn their story, their goals, and their challenges, you demonstrate a kind of leadership that cannot be faked.

If someone is going back to school, ask about their classes. If they are training for a marathon, cheer them on. If they are dealing with a sick parent, offer flexibility. When people know you care about them as a person, not just as a position, they give you their best.

I once worked with an NCO who had risen through the ranks of life the hard way. Life had not given him many handouts, and it showed, not in bitterness, but in how seriously he took his responsibilities, not only at work but also in his role as a father. He was soft-spoken and introverted, the kind of guy who did not say much unless it mattered. But when he worked, you noticed. He was sharp, methodical, and never cut corners.

As I often did with the soldiers I mentored, I asked him one day what success meant to him. He paused for a moment, then said, "I want to be an officer just like you." There was no arrogance in his voice, just quiet certainty.

I told him, "Then go for it. There are only two types of people: those who say and those who do, and you have to decide which one you are going to be."

Without telling anyone, he had been attending college classes at night after duty hours. No fanfare. No complaints. Just quiet progress in the dark hours when others were home or asleep.

A few months later, he walked into my office, a little nervous but determined. "Sir," he asked, "Can I have a four-day pass?" (For the civilian readers, that is essentially paid time off.) Now, I did not hand those out freely; my team was aware of that. But something about the look in his eyes made me pause. There was a tightness in his face like he was holding something in. I could see emotion creeping in around the edges. So I stood up and gently closed the door behind him.

Then he told me why.

He said he had finished his degree and wanted to walk across the stage at his graduation ceremony. He then shared something that stopped me cold: he would be the first person in his entire family ever to earn a college degree.

I did not just say yes. I congratulated him, shook his hand, and told him how proud I was of him. He smiled, but it was the kind of smile you remember: a mixture of pride, relief, and maybe even disbelief that he had made it. He walked that stage.

Later, he applied to Officer Candidate School, and he continues to serve on active duty as a commissioned officer today. And I could not be prouder of him.

That is what happens when someone is seen, heard, and supported. That is the power of believing in people, not just for who they are but for who they are becoming.

Let People Grow

Nothing says "you matter" like investing in someone's future. That might mean cross-training them in a new role. It might mean sending them to a course or just encouraging them to take the lead on a new project.

One of my greatest joys as a leader has never been about the rank I wore or the titles I held; it has been watching the people I mentored surpass me. That is the real reward. Leadership, at its core, is not about keeping people in place; it is about empowering them to achieve their goals. It is about lifting them, helping them grow, and preparing them to lead long after you are gone.

I remember a junior NCO I once worked with; she was sharp. I mean, this young lady had potential dripping off her like sweat after a five-mile run. She was organized, technically sound, and had the kind of attention to detail that cannot be taught. But she had one issue: she did not believe in herself.

It was easy to see in the way she hesitated to speak up in meetings or how she deferred to others even when she had the correct answer. Confidence was not her strength yet.

So, I did what any leader should do when they see someone holding back: I gave her a shot. I asked her to run a small training event. Nothing huge, just a block of instructions on a few basic tasks. She prepared diligently, but when it came time to deliver, she stumbled. Her voice shook, and she lost her place a few times; afterward, I could tell she was beating herself up internally.

I pulled her aside and said something I believe every leader needs to hear early in their journey: "You do not have to be perfect. You just have to keep showing up and keep getting better." And to her credit, she did.

Every few weeks, I gave her a little more responsibility. A larger class. A more complex scenario. A louder room. Each time, she grew a little more. Her posture changed. Her voice grew steadier. She started mentoring others, taking initiative, and sharing ideas in planning sessions.

She went from a young lady who barely spoke to a lady to whom people started looking to for answers.

By the time my assignment ended and I rotated out, there was no doubt in my mind who should take over my responsibilities. It was her. And she did. She stepped into that leadership role and made it her own. That is what authentic leadership looks like.

It is not hoarding the spotlight or gatekeeping opportunities. It is about passing the torch and cheering the loudest when your people run further with it than you ever did.

Years later, I ran into her at a conference. She was now leading a much larger unit, and we caught up over coffee. Before we parted ways, she said, "I still remember the first time you told me I was ready, even when I did not believe it. That changed everything for me."

And that is the thing about value; it is not always found in awards or evaluations. Sometimes, it is in the quiet moments when you tell someone, "You have got this." And they believe you just long enough to prove it to themselves.

Final Thoughts: Value Is a Choice

Creating value in your organization does not require a new title, a bigger budget, or a leadership seminar. It requires intention.

Choose to see people. To hear them. To show up on time. To speak last. To say thank you. None of that is complicated. All of it matters.

People do not just want to be part of a team; they want to be part of a *winning* team. They want to feel like they matter to it. And when they do, it will be evident in their effort, their loyalty, and their performance.

The best organizations are not built on products or profits; they are built on people.

When I retired from the Army, one thing stuck with me more than the medals or titles: the people who said, "Sir, you made me feel like I mattered." *That* is legacy, and that is what creates value.

Chapter Fifteen

The Empty Slice

How Lazy Shows Up

S uccess is not a mystery. It is a pattern. A repeatable process. But most people still miss it, not because they are incapable, but because they choose an easier path. Scientific studies show that our brains are wired to choose the easy option.

Let me introduce you to what I call the "Pie of Success." Imagine a pie chart divided into four slices: The Successful (3%), The Lazy, The Self-Eliminated, and The Uninformed by Choice. Over the next few pages, I will walk you through these slices, starting with the one that swallows the most people: laziness.

- **The Successful (3%)**

- **Lazy**

- **Self-Eliminated**

- **Uninformed by Choice**

Success gets talked about like it is this immense, mysterious treasure that only the lucky or ultra-talented can find. However, the truth is that success is far more predictable and avoidable than most people want to admit. It does not disappear; it gets missed. Not because people are not capable but because they do not do what is required to claim their slice. Some chase comfort, others blame circumstances, and many just never learn how to navigate the game. However, what makes it more real is that I have seen all four types up close. I have lived among the lazy, the scared, the stuck, and, yes, even the successful. Let me tell you about a moment that made it all clear.

Have you ever been roped into a conversation with a bunch of grown adults talking about sports? If you are a fan, you know the routine. But if you are ever bold or foolish enough to ask one of them what sport they played growing up, brace yourself. Get ready for the full story in three parts: First, how they were a hometown hero. Second, how they helped their high school reach the state championship. Third, the usual "I could have played Division I, but..." story. You know, the one with an injury, a jealous coach, or some weird politics that stopped them from making it big. It is practically a rite of passage in every barbershop and backyard BBQ.

Those stories are a perfect example of the Pie of Success. Talent, opportunity, and even passion do not guarantee you a spot in the 3%. Plenty of people start strong but never finish because they settle into one of the other slices.

Now, let me keep it real. I love sports, especially football. But I was not the kid making the highlight reel.

My size and athletic ability placed me solidly in the "you are only on the field because someone had to run down kicks" category. Translation: I was a proud member of the bench mob, only seeing action on special teams. I was not a *bad* kid; I was just *average*. I had the passion, the drive, the hope... but talent did not get the memo.

But I did have a close friend who was the real deal. This guy was a walking highlight tape, effortless on the court and electric on the field. He had the size, the speed, and the skill. The kind of athlete that made you say, "Yes, I went to school with that dude," as if you deserved partial credit for his success. I was convinced we would see him on TV one day.

But talent does not guarantee trajectory. Bad decision after bad decision chipped away at what should have been a golden road to success. My friend did not just fall short; he fell hard. The last time I saw him, he was living on the streets. And that hit me differently because I knew that struggle all too well.

Only 3% of people make it to the successful slice. However, here is the good news: there is room in that slice for more. Getting there does not take a miracle. It takes consistent effort, discipline, and the refusal to live in the other three slices.

Before we get into how to join the 3%, we need to understand why most people do not. Let us start with the most significant slice: laziness.

Lazy Is Not One Thing

Laziness has layers. It is not always the person lying on the couch doing nothing, or the twenty-something in his parents' basement playing video games and drinking their beer. Let us break down the differences in how laziness can appear from different angles:

The Content Lazy

These people have worked hard in the past. They know how to hustle, and they have done so once. However, they reached a goal, became comfortable, and stopped pushing. They mistake one victory for the finish line.

Back in my early Army days, making sergeant as a young soldier was like hitting the jackpot. It was the first real badge of leadership, your first stripe, your first shot at leading soldiers, and your first taste of responsibility that could not be passed off to someone else. If you had strong leadership and stayed locked in, you could achieve that promotion in just two to three years, depending on your military job (MOS). The promotion system operated on a points system, and the Army issued monthly cutoff scores, much like lottery numbers. You could excel in every area, but if your points did not hit the magic number, you were stuck waiting until the next month.

I was performing well. My PT was solid, my rifle qualification was strong, I had college credits, correspondence courses, you name it. However, month after month, I was just shy. On one occasion, I missed the cutoff by two points. *Two points.* It was as if the Army was dangling a carrot just out of reach. Finally, a few months later, the points dropped, and I achieved the rank of sergeant. I was elated. I had been in the Army for only three years, and I had already reached my goal. I felt as if I were walking on air, like the NFL had just drafted me.

Then... I hit cruise control. I did not even realize it at the time, but I became what I now call "content lazy." I was not slacking off, but I was not pursuing growth either. Then it happened: I ran into an old battle buddy about four years later. He was already a Staff Sergeant and preparing for Sergeant First Class. Meanwhile, I had gained weight and slowed down, and my PT scores had quietly become a liability.

That was my wake-up call. I had won the sprint, but I forgot life was a marathon. It snapped me back into focus, and I got to work. Promotions do not crown you a leader; they give you the platform to become one.

If you stop growing after the title, you were not chasing leadership; you were chasing rank.

The No-Knowledge Lazy

That is the Content Lazy. However, laziness has another face, the one that hides behind ignorance. Let me introduce you to Cedric. This is the group that does not read the company manual. They do not seek knowledge, and they associate with others who are equally stagnant. They are not unintelligent; they are willfully uninformed. They do not know how to grow, and, worse, they do not care to learn.

One of my first jobs after transitioning out of the military was with a major retail company. That is where I met Cedric, a warehouse technician who moved with quiet precision and a no-nonsense work ethic. He was not flashy, nor was he loud, but everyone on the floor knew: if something was broken, Cedric did not make excuses; he ensured it got fixed. If someone fell behind, Cedric was there, lifting more than his share of the load. He arrived early, stayed late, and never missed a day.

Yet Cedric had been in that same role for seventeen years. Curious, I asked him one day, "Cedric, have you ever applied for an Area Manager position?" He looked up from his paperwork and said simply, "I have. I was never selected." I followed up, "Has anyone ever assisted you with your résumé? Reviewed it? Prepared you for interviews?" He shook his head. "No one ever offered."

He was not bitter. He was not defensive. He was stating a matter of fact.

That is when it hit me: Cedric was not stuck because he lacked potential; he was stuck, because no one had ever shown him the system. He thought promotions came when someone *noticed* him. He did not realize that, more often than not, leadership roles do not go to the loudest or the longest-serving; they go to those who learn how to *navigate*.

I sat down with him. We reviewed how to craft a résumé that effectively conveys his skills and experience. I showed him how to prepare for interviews, how to read between the lines on internal job postings, and, most importantly, how to advocate for himself.

Cedric had performed for seventeen years of heavy lifting without ever raising his hand to ask for help. Not because he did not want more, but because he did not know how or what to ask.

Now that he has the tools, I can only hope he continues, because someone like Cedric does not need to become a leader. He already is. He just needed someone to help him *translate his work into words that the system could recognize.*

Hard work is admirable. However, if it is not *seen, guided,* or *aligned*, it often goes unrewarded. If no one taught you the rules, you are not behind; you have just not been coached yet.

The Complaining Lazy

They always have a reason why they cannot succeed, and it is never their fault. They blame race, gender, their boss, and the system. While bias exists in real ways, this group relies on it as a crutch rather than a motivator.

While working as a staff member, I stood out, not only because I was one of the few African American officers in the building, but also because I quickly became the unofficial counselor for soldiers who shared my background. Senior NCOs would stop by my office regularly, not because I could pull strings, but because I was a neutral figure, a safe space. I was not in their rating chain, and I was not going to sugarcoat the truth.

One day, a staff sergeant walked in, visibly frustrated and ready to vent. He did not make the promotion list, and he was furious about it.

According to him, it was *everyone's* fault. His leadership was holding him back. The system was broken. The board was biased. The alignment of the stars? Probably off that week. He blamed everyone but Jesus and himself.

Then I asked a simple question: "How many college classes?"

"None."

"Well," I said, "you came here for my opinion, so I am going to give it to you straight. You have not done the work to be competitive. You want the rank, but you have not touched the things that get you there." He blinked, stunned. It was as if I had just dropped a new language on him.

I leaned in and said, "Look, you are not being held back; you are standing still. And if you do not change course, college, special duty, professional development, you will be sitting at this same desk next year, having the same conversation." Wanting success is not the same as working for it. Excuses do not stop progress; they stop you.

The Lack-of-Confidence Lazy

They know what it takes, but they do not believe they have it. They psych themselves out. If the next role requires a certification, they assume they are not smart enough to earn it.

I once mentored a young sergeant who was sharp, detail-driven, and a natural-born leader. You know the type, the kind of soldier you would trust with a mission and sleep easily knowing it is handled. When the opportunity arose for him to apply for OCS, I thought it was a no-brainer. However, he hesitated.

As the talkative person I am, I asked him, "What are you waiting for?"

He shrugged and said, "I do not think I am ready."

"Why not?" I pressed.

"There are better and smarter people for this. I have just been lucky to get this far."

That is when I stopped him. "Lucky? No, luck is simply complex, hard work that people do not recognize. You have been showing up every day, leading, solving problems, and earning trust. That is not luck. That is leadership."

But I realized the issue was not readiness. It was *worth*. Somewhere along the way, doubt had convinced him that growth and recognition were something that *other* people received, not him. People with better résumés. People, without a doubt. So, I gave him an assignment:

"Write your own recommendation letter for OCS. Pretend you are the mentor. Brag on yourself, but be honest."

He returned two days later, holding that letter as if it were a mixture of 100 pounds of fear and hope. I read it. And man, it was all there, the discipline, the grit, the kind of steady leadership that makes others stronger.

I looked at him and asked, "If someone else handed this to you, would you think they earned a shot?"

"Of course," he said.

"Then stop acting like you do not."

He applied. Got accepted. And today? He is mentoring others who are exactly where he once stood, capable but uncertain, qualified but questioning.

Sometimes, the only thing standing between you and your next level is not skill or timing; it is the story you have been telling yourself. Change the story. Change the outcome. You are not here by accident. You are here because you have earned it; now act like it.

The Ignorant Lazy

They think they have made it, but their success was years ago. They have not grown since. They are still riding the high of a win from a decade ago. Progress stopped, but they never noticed.

Once, I worked with a Staff Sergeant, what we call a mid-level leader in the Army. On paper, he was everything the military asked for: promoted ahead of schedule, sharp, uniform in great shape, and could quote regulations word for word. He earned his first promotion fast and hit the ground running.

But somewhere along the way, he stopped moving forward.

He was still leading as if it were a decade ago, relying on outdated policies, resisting change, and avoiding new challenges. When the unit rolled out a new physical training program designed to improve performance and reduce injuries, he pushed back: "I have been doing physical training (PT) this way for years. Why change now?"

PT is the Army's version of your morning team workout. It is how we measure not just fitness but also discipline, leadership, and adaptability.

At first, his resistance was annoying. But eventually, the impact showed.

One of my junior sergeants, a younger team leader, pulled me aside and said, "Sir, can I be honest? Our squad leader is not helping us grow. He shuts down our ideas and refuses to implement the new program."

That was tough to hear about someone who had once been a rising star. But it was true. He had gotten promoted and then stopped earning it. I called him in. No lecture. No rank-pulling. Just one question: "When was the last time you grew as a leader, not in title, but in substance?" He did not have an answer. That silence said it all. The rank is rented. Growth is earned. Titles look good on a résumé, but leadership is proven daily. When a leader stops growing, they do not just stall; they drag their team down with them.

The Real Reason People Fail

Here is the truth: most people are not willing to put in the work. People claim to want success, but we do not act as if we genuinely do. It is like saying you want to look like an athlete while eating like a toddler. Wanting results does not mean much if you are not doing what it takes to earn them.

I once saw a video of a pastor that stuck with me, not because of a sermon but because of a simple, powerful principle: *take your goals with you wherever you go.*

In the video, the pastor was speaking about the Bible, but the principle applies to various aspects of life, including faith, fitness, finances, leadership, and so on. To make his point, he called on a man from the congregation who looked like he belonged in the NFL as a linebacker, built like a brick wall.

The pastor smiled and said, "This man and I go to the same gym. We lift at the same time. We use the same equipment. So why do I not look like him?" The room laughed.

Then he dropped the real message: "Because *he takes the gym with him wherever he goes.* I leave the gym and grab a fast-food breakfast sandwich. He leaves and grabs a smoothie. When I am looking for a cheat meal, he stays focused. He does not *just work out*; he *carries the purpose* behind the workout with him."

That hit me hard. Because it is not about what you do for one hour in the gym, at work, or in your journal; it is about what you do with the other twenty-three hours. If your goals are not aligned with your choices, habits, and discipline, then you are just visiting them.

It is why the fitness analogy fits so well. You see a bodybuilder or athlete and think, *I wish I looked like that.* But while you are reaching for chips, they are reaching for carrot sticks. They are in the gym when you are watching TV. That difference is where success is built.

Promotion Is Earned Before It Is Given

Let us bring it closer to home. You want that promotion that everyone knows is opening soon. But here is how you show up: two minutes before your shift starts, saying, "I am not late." You offer no ideas during team meetings. You never volunteer to help coworkers through problems. You raise concerns but never provide solutions.

In the military, we say during promotion ceremonies: "You are promoted not for what you are doing, but for what you are expected to accomplish in the future."

If you cannot handle your current responsibilities with excellence, why would your organization trust you with more? If you have ever said, "That is not in my job description," it may not be, but do not think for one second that the people responsible for selecting the person for promotion will not remember either.

Final Thoughts

If you saw yourself in any of these lazy categories, that is okay. This chapter is not written to judge you; it is written to scream from the mountaintops: "Wake up!" We are sometimes in a position and may never realize it. There is room in the 3% slice of success. But you will not fall into it by accident. You have to choose it.

Success does not avoid people. It just does not hang around those making excuses. Which slice of the pie have you been living in? What would change if you acted like someone in the 3% today?

Success is not complicated. It is just hard, and hard is where the rewards live.

Chapter Sixteen

Preparation Over Panic

A military phrase we learned from day one was, "Under stress, you will do what you are trained to do." Before every airborne operation, soldiers go through the same ritual: **pre-jump training**. It does not matter if it is your first jump or your fifty-first; every detail is rehearsed: emergency procedures. Malfunction drills. Proper exits. Landings. Every step is covered, every time. Why? Because when the doors of that aircraft open at 1,250 feet and your body floods with adrenaline, you are not going to rely on your talent or your intelligence. You will fall back on your training.

Just like soldiers rely on drilled responses when adrenaline hits, leaders in any high-stakes environment fall back on the habits and preparation they have built long before the crisis arrives. Stress does not create skill; it reveals what you have practiced.

It is no different in leadership. Under stress, in moments of conflict, crisis, or uncertainty, **we do not rise to the level of the challenge. We fall to the level of our preparation.** That is why leadership is not about what you say in the spotlight. It is about the habits, systems, and disciplines you build when no one is watching.

When you are under stress, your brain does not operate the same way it does in calm, controlled situations. It switches from **deliberate, rational thinking** to **instinctive, reactive thinking**, from the **prefrontal cortex** (the center of logic and decision-making) to the **amygdala** (the center of emotion and survival). This shift is fast. It is powerful. And if you are not prepared, it can be dangerous.

That is why we freeze, or lash out, or default to the easiest, most familiar response, regardless of its effectiveness.

In the military, we knew this well. That is why we did not just *know* what to do; we trained for *it* until it became muscle memory. Because under stress, **you will not rise to the occasion; you will fall to your level of preparation.**

The same is true in leadership.

When pressure hits, when the deal falls through, the project fails, or your team starts falling apart, your brain is not going to calmly weigh pros and cons as it does in a classroom. It is going to look for shortcuts. And in that moment, **the only thing that saves you is what you have trained for.**

Training in the Real World

Just as a soldier cannot improvise during a jump, corporate teams cannot rely on instinct alone. The moments of stress in the office, such as presentations, looming deadlines, and high-stakes clients, require rehearsal, preparation, and the muscle memory of disciplined habits.

Stress shows up in corporate life in more ways than you can count, and there is no simple formula to predict or fix it. Maybe you have to present to a senior vice president, and you are not a confident public speaker. Maybe your team is already running on fumes, and the deadline is still looming. Or perhaps you have just received pointed feedback from your boss, and now they are dropping by to check your progress. These moments and countless others can trigger a stress response.

The truth is, stress in itself is not the enemy. The real danger is being unprepared for it. And preparation is not just "knowing what to do" in theory; it is training for what happens when pressure hits.

Let us be honest: most training programs are designed for ideal conditions, not the reality of pressure, confusion, or conflict. In the real world, things go wrong. Communication breaks down. Plans collapse. Emotions run high. That is precisely why training must be deliberate, repetitive, and realistic, because stress does not care about your PowerPoint slides or mission statements.

In the Army, we did not train until we got it right; we trained until *we could not get it wrong*. Repetition under stress builds what we call "muscle memory," but it is mental wiring. You are carving neural pathways that your brain can follow when panic hits. That is how a squad leader clears a building under fire. That is how an executive keeps their composure when the stock tanks, a client walks, or the media starts calling.

Here is the kicker: you do not rise to the occasion; you fall back on your training. And most people, most teams, have not trained for the moment that matters most.

Real-world Training Truths

- **Practicing under pressure:** role-play the hard conversations, run the "what if" drills, and simulate the challenging moments so your people learn to move through discomfort, not avoid it.

- **Making decisions with incomplete information:** life rarely hands you the whole picture. Train your team to act, not freeze.

- **Debriefing every mistake:** after-action reviews are not only for the military. Review, refine, and repeat until the right behaviors are second nature.

Training must go beyond knowledge; it must create habits that stick when your people are tired, frustrated, afraid, or overwhelmed. Because when the heat is on, people do not default to what they know; they default to what they have done.

If your team has not been tested in practice, do not expect them to perform under pressure. And if they fail, that is not on them. It is on the leader who never prepared them.

Emergency Treatment

The principle is not limited to boardrooms or battlefields. Life-and-death situations follow the same rules: preparation determines response. My wife Kathy, an Army nurse who spent twenty years in the Army, experienced this firsthand, so I once asked her to share two moments from her career: one in which training kicked in during chaos, and one where the lack of training made a difference.

At the time, she was working in a non-emergency clinic when she heard a commotion coming from a treatment room. Stepping inside, she saw a patient struggling to breathe. The medics and new doctors on the scene were trying their best, but their approach was unlikely to succeed.

In an instant, her training took over. She recognized the signs of anaphylactic shock, a life-threatening allergic reaction, and knew there was no time to waste. She took command, administered an EpiPen, and within moments, the patient's breathing had returned to normal.

Later, during the after-action review, it became clear that more emergency response training was needed. In that moment, the difference between a good outcome and a tragedy was not luck. It was preparation, confidence, and the ability to act under pressure.

When stress hits, you do not rise to the occasion; you fall back to the level of your training.

The Miracle on the Hudson (2009)

From a single patient to hundreds of lives, the principle remains the same: when adrenaline spikes, you fall back on preparation. No moment is too small or too large for realistic rehearsal. Shortly after takeoff from New York's La-Guardia Airport, the Airbus A320 struck a flock of geese, causing both engines to fail. Captain Chesley "Sully" Sullenberger and First Officer Jeffrey Skiles had spent years drilling emergency scenarios in simulators, including engine-out procedures. Their training allowed them to remain calm, communicate clearly, and make rapid decisions under intense pressure. Instead of trying to return to LaGuardia and risking a crash into the city, Sully chose to land the plane in the Hudson River. All 155 passengers and crew survived.[1] It was not just quick thinking; it was the *muscle memory* built from consistent, realistic training. The crew executed as if it were just another drill, even though lives were on the line.

---

1. "Miracle on the Hudson: US Airways Flight 1549 Lands Safely on River," *The New York Times*, January 16, 2009, https://www.nytimes.com/2009/01/16/nyregion/16plane.html.

What Training Looks Like in Leadership

There is a big difference between **knowing** something and being **prepared** to act when it matters. Knowledge is information you store. Preparation is the muscle memory that kicks in when stress tries to overwhelm you.

I learned this lesson the hard way myself. The stakes are real, the risk is yours to manage, and there is no room for improvisation. Let me take you to my own airborne experience. I was the Assistant Jumpmaster on an airborne operation just outside Aviano Air Force Base in Italy. The Jumpmaster was responsible for inspecting every piece of equipment, checking each paratrooper's gear twice, and rehearsing every step of the process. In the aircraft, we were responsible for timing the drop, ensuring every paratrooper exited in the correct order, and reacting instantly to anything that goes wrong. I started exiting all the paratroopers on my chalk, and after watching my last jumper exit, I prepared to follow. But the moment I stepped out, I realized I had made a bad exit and ended up completely upside down. My parachute deployed, but my risers had wrapped tightly around my left leg.

I did not have time to think through a step-by-step checklist. Training took over. I swung my body in a full circle to untangle the risers. On the last twist, I felt a sharp *pop* of pain shoot through my leg, and I was certain it was broken. But the mission was not over. I still had to land, and I knew the only way to avoid disaster was to focus on slowing my descent and controlling my landing.

Later, I would learn I had torn my ACL. Painful? Yes. But I was alive, and that was because the training I had repeated countless times had become instinct.

Most people think knowing the correct answer is the same as being ready for the challenge. It is not. **Knowledge** is understanding what should be done; **preparation** is building the skill and resilience to do it, especially when the pressure is on. Just as a pilot does not "learn to fly" during a thunderstorm, a leader cannot expect to learn how to lead in the middle of a crisis.

Leaders who train well focus on real-world challenges long before they arrive.

- **Practicing difficult conversations**: Conflict avoidance is a sure-fire way to let minor problems grow into big ones. Leaders rehearse tough talks, role-play scenarios, and develop the confidence to speak truth with empathy and clarity.

- **Setting clear expectations**: Ambiguity creates confusion. Training means learning to communicate goals, responsibilities, and boundaries so there is no room for guesswork.

- **Delegating before there is a crisis**: Too many leaders wait until they are overwhelmed before handing off responsibility. Prepared leaders identify and empower others early, so when the heat is on, the team already knows what to do.

- **Building trust daily**: Trust is not built in the storm; it is built in the calm. Leaders invest in relationships, show consistency, and prove they have their team's back, so that in challenging moments, people follow with confidence.

- **Learning how to pause and respond instead of react**: Under stress, untrained leaders tend to act on impulse. Trained leaders create the habit of slowing down just enough to choose the best action instead of the quickest one.

In leadership, training is not about memorizing policies; it is about turning effective behaviors into muscle memory. That way, when the moment of truth arrives, your actions are automatic, confident, and aligned with your values.

What Happens Without Training, or Worse, Overconfidence?

As I mentioned earlier, I was qualified as a Jumpmaster. On this particular jump, several people from my section were on my aircraft. As a courtesy, I asked each one whether their rucksacks were rigged to be lowered adequately. Everyone answered with a confident, "Roger that, sir."

I then approached my platoon leader. This was only his sixth jump, and his first since graduating from Airborne School, so I asked him personally if he was ready to go. With a pompous tone, he replied, "Sergeant, I know how to rig my rucksack."

I simply responded, "Roger that, sir."

We completed the manifest call (a term used for attendance), and the jumpers then began their actions in the aircraft. I stayed behind with the other Jumpmaster, who happened to be our battalion sergeant major, inspecting rucksacks. Within seconds, the CSM was shouting, tearing into one rucksack, and ripping it apart.

The rucksack belonged to my platoon leader, and the sergeant major tore into me just as hard. I had to pick up the pieces, correct the problem, and ensure the jump could proceed safely.

Preparation is not only about your own skills; it extends to those you lead. Confidence without training can endanger the team, as I saw firsthand. Was this failure because my PL was not trained? Or because he was not prepared? The truth is it was both. His lack of experience, combined with an unwillingness to ask for help, created a risk in a moment when flawless execution was the only thing that mattered.

In leadership, whether in the military or business, confidence without preparation is dangerous. A leader who refuses to ask for help risks not only their success but also the safety, trust, and performance of the entire team. Authentic leadership means being humble enough to double-check, willing enough to learn, and wise enough to prepare long before the moment of execution arrives.

Building a Training Culture

Once you see how preparation saves lives and outcomes, it is clear: training is not optional; it must be embedded in the culture. An authentic training culture does not happen by accident; it is built intentionally by leaders who model the mindset themselves. Leaders must teach their teams that training is not a one-and-done event; it is a continuous cycle of learning, practicing, evaluating, and improving. Build this mindset into your team.

- **Repetition without complacency**: Practice often, but keep each session intentional so familiarity never becomes carelessness.

- **Simulations, role-plays, and rehearsals**: Recreate the real-world environment so your people can learn under realistic conditions, not just in theory.

- **After-Action Reviews (AARs)**: Discuss what worked, what did not, and what changes need to happen.

- **Accountability systems**: Ensure people own their roles and are measured against clear expectations.

Training must be embedded into the fabric of your organization. It has to be a priority; otherwise, you will always find an excuse not to do it. There will always be more pressing things that come up. But consider the alternative: How would you explain to a family, *"I am so sorry your loved one died because we did not practice something as simple as a fire drill on a night shift; it would have been too difficult"*?

In the end, the stakes are too high to treat training as optional. A culture that trains consistently will execute confidently because when the moment comes, the team will not have to rise to the occasion. They will fall back on their preparation.

The paratroopers are at the door. The red light is about to hit green. The engines are roaring, and your people will fall back on their training. When the doors open on your next high-pressure moment, what will they fall back upon? Do not wait for the fear and the adrenaline to highlight the gaps. Identify a skill scenario or decision you have not trained for, then commit to practicing it this week. The time to prepare is not when the plane is over the drop zone. It is right now! When you are standing in the door, there is no going back. Remember: under stress, you will do what you are trained to do. Ensure that your training today prepares you for your response tomorrow.

Chapter Seventeen

The Man in the Mirror

Leading Yourself First

Have you ever gone to a restaurant and received bad service? Maybe the waiter ignored your table, delivered your order incorrectly, or moved at a glacial pace. How long did it take for you to decide you were dissatisfied? Seconds. Some of us will not even wait until the meal is over before asking for a manager.

Or maybe you have had your luggage lost at the airport. You filed a claim for the missing bag and a complaint for the poor customer service. Understandably so. When we do not receive the service we expect, we let someone know about it quickly. Now here comes the million-dollar question: Would you hire yourself to achieve the goals you have set for your life?

We are quick to hold others accountable. We expect results. We expect timeliness. We expect consistency. And yet, when it comes to our own lives, our habits, our follow-through, our discipline, we often let ourselves off the hook.

Imagine applying the same standard to yourself that you apply to others. If you showed up for your own dreams the way that waiter showed up to your table, would you leave a tip, or would you complain? Would you say, "That is someone I trust with my future?"

In Michael Jackson's song "Man in the Mirror," he sings, "I'm starting with the man in the mirror. I'm asking him to change his ways."[1] It is a powerful reminder, but one we rarely act on. It is far easier to blame than to take ownership, easier to complain than to correct.

Real leadership begins with self-leadership. Self-leadership begins with self-honesty.

So here is the challenge: Before you critique others, ask yourself a more complicated question. *Would I myself? Would I trust myself? Would I be inspired by myself?*

Your goals, your dreams, your leadership, they all start with the person staring back at you in the mirror. Let me tell you how I started my Army career. That realization did not hit me in a classroom or a conference room. It came in uniform, early in my Army career, when I first discovered how easy it is to hold others accountable but let *yourself* off the hook.

I started my military career as an **E-1**, the very first enlisted rank, similar to being an entry-level employee with no experience. To get promoted to **E-5**, better known as **Sergeant**, you do not just wait *for* your turn. *You earn it.* The Army makes you fight for it.

As I mentioned in a previous chapter, it was not until I ran into someone I knew who was surpassing me that I knew I needed to look at the man in the mirror. He had that fire in his eyes, and I simply lost mine. That was when it hit me: titles do not change habits. Promotion does not guarantee passion. What got me to Sergeant was not enough to keep me growing. The same truth also applies to business and life. He reminded me of the version of myself I used to be. That was when I realized something that applies far beyond the military: **What got you here will not get you there. You have to lead yourself first!**

Just like in the corporate world, reaching the next level is not just about performance. It is about consistency. Self-leadership is about staying hungry after the title arrives.

The Core Traits of Self-Leadership

---

1. Michael Jackson, "Man in the Mirror," track 8 on *Bad*, Epic Records, 1987.

Real leadership starts with leading yourself. That means being brutally honest about whether you are doing the work when no one is watching. Over the years, both in uniform and in business, I have found that self-leadership comes down to three simple traits:

Consistency

Consistency is not only about the big moves; *it is* about making the right moves. In the Army, no one had to tell you to lace up your boots. During my time, we had to shine our boots, and we knew that arriving at 9:00 a.m. for a 9:00 a.m. formation was considered late. I once worked with a soldier who struggled with physical fitness, specifically running. She would work on this after work, and even when we were deployed, she would find a way to walk/run around the FOB. Leaders saw it, but more importantly, her peers saw it. Before you could blink an eye, she was promoted before people with more experience than she had. You do the small things every day, rain or shine. The same applies in corporate life. When I joined a new corporate team, I was told that an associate was the person the leadership team could trust. As a new team member, if I had a question, he was the person I would go to, and he came through every time, never late. His reports were immaculate, and he was also the person we used to run our onboarding program. He, too, ended up getting promoted and doing well for himself.

That is the thing about consistency: it is rarely flashy, but it is always foundational. It is not about showing up for the one big event; it is about showing up the right way, again and again. Whether you are a junior soldier working on your physical fitness or a seasoned professional managing onboarding in a high-performing team, the principle is the same: trust is built in repetition.

Consistency earns credibility, and credibility is what gets you promoted, both in rank and in influence. So if you want to lead others, start by showing you can lead yourself: on the good days, the hard days, and the ordinary days in between.

Because in leadership, reliability and consistency go hand in hand, it is a superpower. And those who master the small things daily are the ones who end up trusted with the big things tomorrow. But consistency alone is not enough.

Even the most reliable routines are tested when the conditions are hard. That is where discipline shows up.

Discipline: Do You Show up on Hard Days?

Motivation is overrated. It comes and goes. What matters is what remains when motivation runs out, and that is called discipline. Discipline is the gap between where you are and where you want to be. Anyone can perform when conditions are perfect. But real leaders show up when it is hard.

During my early days in the medical field, we did not just train to treat trauma; we trained to do it under pressure. Bad weather. Middle of the night. Back of a helicopter. No clinic. No comfort. Just skill, grit, and discipline.

Even simple procedures became difficult in those environments. But that is the point. So let me say one of my favorite phrases. *You do not rise to the occasion; you fall to your level of preparation.* And if you have disciplined yourself to perform when it is uncomfortable, you will be ready when that moment comes.

Push yourself to do the hard things because they *are* hard. You will be surprised how much easier the future feels when you have already trained for adversity. Discipline is not about how you feel; it is about what you value.

Anyone can lead when conditions are ideal. The ones who rise when everything is off balance set the standard. So do not train for when it feels good; train for when it does not. Still, consistency and discipline without humility can turn into arrogance. And arrogance is the enemy of growth. That is why the third trait matters.

Humility: Can You Admit When You Need to Grow?

When Satya Nadella became CEO of Microsoft in 2014, the company was seen as outdated, bureaucratic, and losing its edge in innovation. Previous leadership had focused heavily on competition, especially trying to beat Apple and Google, rather than on learning, listening, and evolving.

Nadella came in with a different posture. He did not pretend to have all the answers. In fact, one of his first major internal messages to the company was about the power of having a **"learn-it-all" culture instead of a "know-it-all" culture.** He began asking questions, not issuing mandates.

He focused on listening to employees, customers, and even competitors. That humility set the tone for a complete cultural transformation. Microsoft shifted from internal silos and ego-driven strategies to a culture of collaboration, curiosity, and continuous learning.

The result: Under Nadella's humble leadership, Microsoft's market cap skyrocketed from **$300 billion to over $2.5 trillion,** and the company became known not just for its software but for its empathy, agility, and innovation.

Satya Nadella's story proves it. The most effective leaders are not the loudest in the room. They are the ones asking questions, admitting when they do not know, and creating space for others to contribute. Humility is not a soft skill; it is a strategic advantage.

These three traits: **Consistency, Discipline, and Humility,** are not just *nice to have;* they form the baseline before you lead others, the questions you must first ask yourself.

- Am I consistent in what I do?

- Do I show up even when it is hard?

- Do I own my blind spots and seek growth?

If the answer is no, good. That means you have got a place to start. And that is where all outstanding leadership begins, with the person in the mirror.

Common Excuses That Sabotage Self-Leadership

If consistency, discipline, and humility are the fuel of self-leadership, excuses are the leaks in the tank. They drain momentum before it even reaches the road. Let us be honest; most of us do not lose momentum because of outside obstacles. We lose it because of the excuses we quietly start to believe. I have encountered common excuses over the years.

"I do not have time."

We have all said it. But the issue is not time; it is **priority**. The same people who say they do not have time will binge-watch a show, scroll for an hour, or take on tasks that do not serve their goals. Time is the one thing we all get in equal measure. Leaders do not find time; they make time. If it matters, it gets scheduled. If it does not, it gets cut. So, ask yourself: *What is your priority?*

"I will start after this next thing."

Procrastination always disguises itself as preparation. We convince ourselves we need just a little more planning, the *perfect* window, or the right energy. But the truth is, **there will always be something else.** The discipline to start in the middle of the mess is what separates real leaders from those just waiting for ideal conditions.

"It is not that serious yet."

This one is dangerous. We ignore early signs, burnout, disorganization, and drifting goals because they are not urgent. But leadership is not about reacting to fires. It is about preventing them. The most effective leaders address minor issues early. They do not wait until "serious" becomes a crisis.

"This is just how I am."

That is not self-awareness; it is self-limiting. Attitude is a choice, so choose a good one. Growth starts when we stop defending our dysfunction. Saying "this is just how I am" becomes permission to stop evolving. Real leaders do not make peace with patterns that hold them back. They challenge them and develop a plan to change them.

Excuses may feel comfortable, but they hinder your progress. Self-leadership means recognizing these mental shortcuts and choosing a different path. It is not about being perfect. It is about being honest. The moment you stop believing your own excuses, your growth accelerates.

- What excuse do you catch yourself using most often?

- Where has that excuse already cost you time, trust, opportunity, or growth?

- What would change if you stopped giving that excuse power?

Action Step

Write down the excuse that shows up most in your life. Then rewrite it into a truth.

Example

- Excuse: *"I do not have time."*

- Truth: *"I have not made this a priority, yet."*

Keep that truth visible on a mirror, your phone screen, or your desk, and let it call you up when excuses try to creep back in.

At the end of the day, leadership does not start in the boardroom, the formation, or the conference stage. It begins in the mirror. The reflection staring back is not just who you are today; it is who you are becoming tomorrow. Lead that person first, and you will be ready to lead anyone.

Sit down with yourself and write down the answers to these questions:

1. *Where are you currently avoiding the mirror?*

2. *What part of your leadership, habits, or mindset are you hoping no one notices, but you know requires improvement?*

3. *Are you leading others better than you are leading yourself?*

4. *If your team mirrored your discipline, attitude, or focus, what would you see?*

5. *What excuse are you still protecting?*

6. *Be honest: What is the story you have been telling yourself that is keeping you stuck?*

7. *What would it look like for you to take radical ownership today?*

8. *Not next quarter. Not next year. Today.*

9. *Do you have the courage to be vulnerable in front of your team?*

10. *Who needs to see the human side of your leadership, rather than the polished version?*

11. *What small, consistent change would create the most significant differ-*

ence in your self-leadership?

12. *Forget overhauls. What is your one-degree shift?*

Chapter Eighteen

What Gets Checked Gets Done

H ave you ever watched *Undercover Boss*? A CEO trades the boardroom for the break room, and suddenly they see the truth that policies on paper do not always match what is happening on the ground. It is a vivid reminder that guidance without follow-up is nothing more than noise.

It is the same in parenting. Tell your children to make their beds, but never check, and eventually it will not happen. Leadership is no different. Accountability is not about distrust; it is about reinforcing what matters. And nowhere did I learn that lesson more clearly than during a deployment to Afghanistan.

While deployed to Afghanistan in 2009, I was still a new platoon leader, learning everything I could. I was assigned to the Brigade Special Troops Battalion (BSTB). A BSTB is a multifunctional battalion that supported a Brigade Combat Team (BCT) with specialized capabilities, typically including engineers, military intelligence, signal (communications), and my portion of the mission, military police (MP). The BSTB centralizes these support functions under one battalion to enhance coordination and provide the brigade with critical combat support and combat service support. The bottom line is that we are the support arm for the BCT. We did not own land within the battle space, but due to a new mission, the BSTB would become landowners.

One of our first missions as landowners would be something we never thought we would accomplish during this deployment. It was to conduct a nighttime air-assault mission. We were not infantry, we were not a Special Operations team, we were a military police platoon. Our portion of the battalion's mission was to establish a blocking position while other elements conducted a cordon and search for Taliban fighters. To break this down into standard English for non-military readers, air-assault operations involve flying into the objective by Chinook helicopter. Our task was to block the enemy's avenue of escape from the village where the cordon-and-search was conducted.

When we received word about the upcoming missions. I had some concerns, as this was uncharted territory for many of my soldiers. Some had never flown in a helicopter or conducted a nighttime air assault mission, but this was exactly why I wanted to become an officer. So it was time to put up or shut up! We conducted rehearsals, including entering and exiting the aircraft, setting up our blocking positions, and conducting sensitive items checks before all major movements. We conducted several of these missions in various locations and generally followed the same rehearsal procedures for all of them.

After several successful missions, our team settled into a rhythm for conducting air assaults. We followed the same procedures for each operation: loading and unloading from helicopters, setting up blocking positions, and conducting checks on sensitive items. When we were preparing for our fourth mission, it felt routine, almost automatic. That is when it happened. A soldier left his nine-millimeter pistol behind on the objective.

After this mission, we flew back to **FOB Shank**, and the work priorities dictated that the first thing we did was clean weapons. I was sitting on my bunk doing just that when my platoon sergeant came in and said, "Sir, we have a problem."

I asked, "What is the problem?"

He replied, "One of the soldiers does not have their nine-millimeter pistol."

My first thought was that it had been left on the helicopter, since I knew we had completed a hands-on check before leaving the objective, or at least, that is what I believed. I finished my Rip It, shook my head, and said, "You have got to be kidding me." The golden rule in the Army is that *bad news does not get better with time.* Mac and a few others returned to the airfield to look, and I headed off to see my company commander. While unhappy, Captain Thom remained calm; he had always been. Cool, collected, and level-headed, he was the kind of leader you wanted beside you when everything hit the fan.

When the team returned from the airfield, they still could not find the weapon. We sat in silence, awaiting guidance from our leadership, and finally, we received the word that we would be going back out. No one argued or made a witty comment. Everyone knew what was at stake, and if they were angry, they did not say it or show it.

A storm of emotions hit me, anger that we were going back into the hornet's nest for a single pistol and that I had failed as a leader; frustration that I had been given the privilege of leading a great team and had let them down; and fear that we would be ambushed, or worse, during the return mission.

Our standard operating procedure required "a sensitive items check," in which junior leaders physically verified the presence of all weapons, ammunition, and communication gear, then reported up the chain, culminating with me. I had received that report and passed it along, confident that everything was accounted for. But it was not. The mistake was mine, not the squad leader's. If I kept my job after this, I knew I would have to reinforce that these checks were done thoroughly, not just pencil-whipped as routine or habit. Complacency had crept in, and I had allowed it to happen.

There is always a heightened sense of awareness during normal convoy operations, but it was magnified tenfold during this operation. I needed to ensure that this operation proceeded smoothly, and ultimately, the goal was to locate the weapon and return safely to FOB Shank. Upon returning to the objective, we implemented a security plan and commenced our search. We searched for several hours but were unsuccessful. We wanted to return to FOB Shank before dark for safety and security reasons, and therefore, we abandoned the search. I knew from the beginning that this operation would be like finding a needle in a haystack, and it was true.

We did not find out until the next day that Mac and I would keep our jobs. I am unsure how they made that decision, but I was grateful.

That day seared a permanent lesson: *what leaders do not check does not get done.* It is not enough to give guidance and assume it is followed; accountability requires action. I had fallen into the trap of trusting routine. The sensitive items check was reported up the chain as complete, and I accepted it at face value. But leadership is not about trusting that something was done; it is about verifying it was done *correctly.* Systems and procedures mean nothing without discipline and follow-through. I realized I had failed not because I did not care but because I did not check.

From that point forward, I changed the way I led. I ensured that sensitive item checks were not just check-the-box routines but hands-on verifications. I started randomly spot-checking gear myself, not to micromanage, but to reinforce the standard. I began to mentor junior leaders on *what* to do and *why* we do it. I reminded them that responsibility flows upward, not downward, and that trust must be backed with accountability. Most importantly, I adopted a leadership posture that paired belief in my soldiers with the discipline to confirm that they upheld their duties. Leadership means standing in the gap between failure and success, and on that day, I learned that the gap was filled by *follow-through.*

Takeaway 1: Checking Is Not Micromanagement

All leaders will struggle at some point in their careers with the choice between checking their team's work and micromanaging. It is a safe bet that we all have had a micromanaging supervisor at some point, and we have hated working for them. If the micromanager taught us anything, it is not to be like them. Trust is not the absence of verification; it is the foundation of accountability. Effective leaders do not simply assume that tasks are completed; they confirm. Trust is strengthened, not undermined, when leaders verify that standards are met. It was my fault that I did not enforce the standard requiring checks of sensitive items and ensuring they were secured. Checking is not just about individuals; it is about systems. One leader cannot see everything, which is why the process itself has to hold people accountable.

Takeaway 2: Guidance Without Follow-Up Is Just Noise

Strategic plans, company policies, or executive directives are only effective if accompanied by consistent and intentional follow-up. It is not uncommon for leaders to assume that once expectations are communicated, whether through a memo, meeting, or training session, they will be executed correctly and consistently. But just like in the military, where a missed equipment check can put lives at risk, failing to confirm that guidance is implemented can lead to costly operational failures. Consider the CEO who outlines a new customer service protocol without verifying whether frontline teams are using it, or the compliance officer who assumes everyone is following regulations without conducting an audit. Leaders must move beyond thinking and start verifying. Otherwise, they risk allowing standards to erode under the illusion that *everything is fine*. The organization does not matter; leadership is about ensuring execution and setting direction and climate.

Takeaway 3: Do Not Allow Complacency In

Repetition can produce efficiency, but it also creates a dangerous space for complacency to grow. In business, tasks that once required full attention, such as safety checks, data entry, contract reviews, or client interactions, can become mindless habits over time. When people feel like they have "done it one hundred times," they stop questioning whether they are doing it *right*. Many organizations fall into a false sense of security, skipping the basic steps or assuming others have already covered the bases. That is when costly mistakes happen. For leaders, the warning is clear: just because something *feels* routine does not mean it is risk-free. Good leaders will take the time to put in mechanisms, such as refreshers, spot audits, or process reviews, to halt complacency. When complacency creeps in, accountability does not disappear; it rolls uphill. Ultimately, the leader owns it.

Takeaway 4: Accountability Starts at the Top

In corporate life, it is tempting for leaders to delegate tasks and then detach from the outcomes. As a young leader, I learned early that you can delegate a task, but you cannot delegate the responsibility that comes with it. The responsibility always falls on the leader's shoulders. However, as this story powerfully illustrates, authentic leadership means taking ownership of what happens on your watch, even when others are technically responsible for carrying it out.

Shifting blame onto subordinates is never the correct course of action when things go wrong. Leaders who shift the blame risk eroding team trust and damaging the organization's culture. Leaders who take ownership of mistakes strengthen their credibility. When things go wrong, pointing fingers at your team, or blaming the system, or the organization's *higher-ups*, undermines the organizational culture. Any leader who admits they missed a detail in a project review, or a department head who takes the fall for a miscommunication, sends a louder message than any values statement on the wall. Accountability is not just a leadership trait; it is the foundation for a high-trust, high-performance organization. If failure happens, that is the leader's defining moment. Do you cover it up, push blame, or face it head-on?

Takeaway 5: Failure Is a Crucible for Growth

Every leader will face moments of failure when a decision, or lack of a decision, backfires, a process breaks down, or a team member falls short. This is precisely what occurred on this mission. What defines a great leader is not avoiding these failures but what they do *after* them. In the story, I did not deflect or minimize my mistakes; I reflected, adapted, and improved. It is the same way you should handle matters in business. It does not matter if the new launch of a product fails or an important deadline is missed. It is up to the leader to take ownership of the mistakes. When the leader responds with integrity and courage, these moments become powerful coaching opportunities for future leaders within the organization. Failure can be one of the most effective leadership tools for long-term growth when handled with accountability and transparency.

The lesson is clear: what does not get checked does not get done, plain and simple. But this chapter is not just a story about a lost weapon; it is a call to action for every leader who wants to raise the bar. Accountability is not a sign of mistrust; it is a commitment to organizational and individual excellence. As you ponder the leadership decisions you have made along your leadership journey, ask yourself: Where have I assumed rather than verified? Where has routine become a risk? Use these questions not to assign blame for faults, but to build a better system.

Moving forward, your mission is to create a culture in which discipline is a habit and standards are lived, not simply stated or laminated on a board in the conference room. Trust is strengthened through consistent, respectful follow-up. The real test of leadership is not what you say; it is what you check, what you reinforce, and what you refuse to overlook.

Here are my suggestions.

Set Clear Expectations: Define success from the beginning. Be specific about outcomes, timelines, and priorities, so that your team knows what "right" looks like, especially when you are absent. Clarity upfront reduces the need for over-the-shoulder corrections later.

Delegate Outcomes, Not Just Tasks: Do not just assign duties; assign ownership. Allow them to make changes as they see fit within the correct parameters. This way, they are not simply completing a delegated task; they are completing something they own. When team members are responsible for the *result*, not just the process, they operate with more initiative and take more pride in their work.

Establish a Cadence of Check-Ins: For long-term projects, schedule regular check-ins to discuss progress and issues. Your team will prefer this approach to having someone stand over their shoulders. Weekly check-ins, milestone reviews, and after-action briefings allow you to stay informed without hovering. Regularity beats randomness.

Trust, but Normalize Verification: Verification should be a leadership routine, not a red flag. When checking becomes a standard part of your process, it builds accountability without undermining trust. Use this phrase: "I check because it matters, not because I do not trust you."

Create Psychological Safety: Your team should know it is okay to make mistakes and that their opinions matter. When you create this culture, people will take ownership of the team's projects. Teams that feel safe bring problems forward. Build an organizational culture that values taking charge and learning from failure.

Chapter Conclusion

This chapter could easily end with a single sentence: *What does not get checked does not get done.* However, that truth deserves more than a passing mention; it deserves reflection.

No matter how clear your expectations are, how many times you have said them, or how many Standard Operating Procedures (SOPs) you have written, execution still requires confirmation. This is not because your team is lazy or untrustworthy, but because people, at every level, are human. They forget. They assume. They get comfortable. In high-performing environments, that comfort can quietly drift into complacency.

Leadership is not about doing everything yourself; it is about ensuring that what needs to be done *gets done*, which requires follow-up. Systems help, but systems without leadership are hollow. Processes require reinforcement; standards need to be regularly inspected, and teams need to know that their work matters enough to be reviewed.

The balance between micromanagement and accountability is not as delicate as some make it seem. Micromanagement is driven by fear and control, whereas leadership is driven by purpose and ownership. When your team knows *why* you check and sees that you are consistent and fair, your presence builds trust, not tension.

As a leader, your credibility is built on outcomes, not on intentions. Those outcomes hinge on disciplined follow-through.

Leadership is not about doing everything yourself; it is about ensuring that what needs to be done gets done. Micromanagement is fear; accountability is leadership. The difference is purpose.

And here is the hard truth: if you are not checking, you are not leading. You are hoping. Hope is never a strategy for success in any mission-critical environment.

What gets checked gets done, and what does not, does not.

Chapter Nineteen

React Less, Lead More

The Battle Between Urgent and Priority

When Urgency Meets Leadership

Urgency does not knock; it kicks the door in. Leadership in those moments is not about panic; it is about clarity, courage, and conviction when every second counts.

I was at the gym, finishing a workout and sliding into my flip-flops before heading to the shower. Then I heard a loud thud, followed by someone saying, "I am going to get help."

Curious, I looked around the corner. I expected to see a locker that had fallen or a maintenance issue. What I saw instead stopped me in my tracks: a man had collapsed. He was not moving. People stood around, frozen. Some even said, "Do not touch him." But my training kicked in.

I stepped forward, checked for breathing and a pulse, but there was nothing. I rolled the unconscious man onto his back with the help of a staff member. The staff member was calm and focused, precisely what the situation demanded.

We started CPR. There was no AED on site, so we worked with what we had. It felt like an eternity before the EMTs arrived.

I did not know the man. I did not hesitate. I did not wait for someone to tell me what to do. I just knew this: **doing something was better than doing**

nothing. Waiting for help that might not come was not an option. Getting him breathing and getting him a pulse was the urgent matter at hand.

Months later, while working out, a man approached me and asked, "Are you Greg?"

"I am."

"Do you not recognize me?"

I paused. Then it hit me: he was the man I helped save. We embraced and talked for a few moments. That exchange reminded me of something I have learned time and time again in both uniform and business:

Leadership is not about perfect timing or permission. It is about being present and acting when it matters most.

Most people freeze during moments of urgency, not because they do not care, but because they have not trained themselves to move with purpose under pressure.

Urgent moments will find you, whether in a boardroom, a warehouse, a patrol base, or a locker room.

When they do, your team is not just looking for direction; it is also looking for a clear path forward. They are observing *you.*

Will you panic? Freeze? Wait for someone else? Or will you lead? *That tension, the pull between what feels urgent and what truly matters, is what every leader wrestles with.* Because here is the truth: the more you prioritize what matters *before* a crisis, the more equipped you will be to lead *through* it.

Clarifying the Difference: Urgent vs. Priority

Let us get clear on the definitions of the terms.

Urgent tasks demand immediate attention. They come with deadlines, disruptions, or the weight of someone else's crisis. They press on your chest the moment you walk into the office. They are loud.

Priority tasks, on the other hand, concern long-term importance. They do not always demand attention, but they matter more. These are the tasks that move the mission forward, even if no one checks on them every hour.

Urgency Is About Time. Priority Is About Impact.

You can spend your entire day buried in urgent tasks and still get nowhere meaningful. However, when you lead with priority, real progress begins.

Here is the danger: when everything feels urgent, *nothing* qualifies as a priority. That is how leaders burn out and take their teams with them.

Why Leaders Confuse the Two

Most leaders do not become overwhelmed because they are lazy; they get overwhelmed because they confuse what is loud with what is important.

Urgency and priority both demand attention, but only one deserves it. So why do so many leaders get it wrong? Because urgency feels satisfying. The leader jumps in, solves a problem, and feels useful.

Priority requires discipline. It is not always urgent, but it is always essential. Let us break it down.

The Rush of Being in Demand

There is a rush when someone needs you to solve a problem immediately. You don the superhero cape, step in, fix the issue, and walk away feeling valuable, required, in control, and essential. In that moment, you convince yourself: "This is leadership." However, it is not.

That is an **addiction to urgency**, a problem many leaders have but often fail to recognize. Urgency does not just show up in emergencies; it sneaks into inboxes, meetings, and calendars.

Urgency feels important. Fires give focus. You receive instant credit, immediate feedback, and a quick dopamine hit. You become the hero of the moment. The more you chase that high, the more you ignore the work that builds a team.

My friend and mentor Chris Hodl once told me, "Leadership is not about solving problems; it is about putting systems in place so the problems do not happen." That lesson stuck with me.

You do not get standing ovations for designing systems that prevent those fires in the first place. No one claps when you build infrastructure, align vision, or say "no" to distractions so your team can stay on course. *That work is quiet.* It is often thankless. That is where **real leadership lives**: in the disciplined, strategic work that no one sees until it succeeds.

Urgency is loud and addictive. **Legacy is built in silence**, through the work that feels slow, uncelebrated, and sometimes unnoticed until the team begins to win consistently. If everything feels like a crisis, one may not be leading effectively and may be merely reacting.

The Illusion of Productivity

I became known by my team as the guy who would say, "That sounds like someone else's problem," when colleagues brought me urgent matters that did not fall within my responsibilities. Here is the thing: never allow your top priority to drop to sixth place to address someone else's emergency.

Every urgent situation a colleague brings is *their* top priority. That does not automatically make it yours. And one of the most consistent threats to maintaining focus is email.

Email is one of the biggest distractions affecting many leaders. Half of the emails are not that important, yet we spend valuable time reading the email thread. Here is a method that can help.

Everyone experiences those days: 100 emails, seven Zoom calls, and moving from one issue to the next. You crawl into bed thinking, *I crushed it today.* Then, you pause and realize that nothing meaningful has moved forward.

As a leader, email can feel like quicksand; it pulls you into everyone else's priorities instantly. A leader does not have to allow this. Begin by setting specific times to check email rather than allowing it to control the day. One method changed the game for me.

During an Army leadership training, a senior officer shared a system he used to reduce the noise in his inbox. He required his team to label every subject line with one of the following:

Decision: Use this when you need me to approve something, give the green light, or make a choice. This is high-priority. Get to the point quickly. Do not bury the ask; someone else may be waiting on me.

FYI / Informational: Use this when I do not need to take action, only to remain informed. These are updates I can scan when there is available time.

Review: Use this when you want my eyes on something, maybe a document, report, or draft. It is not urgent unless there is a clear deadline, so provide the timeline explicitly.

This single change helped me serve my organization more efficiently and prevented me from being overwhelmed by my inbox. The truth is this: **If everything is urgent, nothing is.** When I understand what is required, I can provide it more efficiently.

Are you spending your day chasing fires or building something that does not burn so easily?

Fear of Disappointing Others

Many leaders confuse urgency with priority because they dislike saying "no." Leaders care about their teams and want to be available, supportive, and approachable. Consequently, they say yes to tasks that are not aligned with their actual responsibilities, thereby crowding out what truly matters.

But here is the truth: **leadership is not about being liked, as mentioned in a previous chapter. It is about being effective.**

Saying "no" does not make you selfish. It demonstrates focus. Often, it is not even a firm "no", it is a *"not right now."*

A junior manager once approached me in the middle of a deadline week, requesting assistance because one of my associates was dual-trained in his area. The manager explained why they needed help, and it was a good idea; I genuinely wanted to support it. However, I was already deeply engaged in quarterly reporting and preparing for a site inspection. I explained: I want to provide assistance, but not today. I told him personnel numbers will improve tomorrow, at which time I will be able to provide support. This approach allowed me to proceed without increasing risk to myself or the team, which was already short-staffed. He respected this decision. I maintained my priorities and honored his request. This demonstrates balance.

You do not have to ignore people to lead well.

You have to learn to disappoint the right people, at the right time, for the right reasons. That is how you stay clear and effective.

The Consequences of Leading by Urgency

Identifying what to do is one thing. However, if you do not change how you lead and stay stuck in a reactive mode, there are real consequences for your entire organization.

Urgency is sometimes necessary. However, when it becomes the default rather than the exception, everything begins to suffer: mission, morale, and momentum.

Mission Drift

Chasing Calls or Leading the Mission?

When I commanded a military police section, our mission was clear: maintain order, ensure safety, and support force protection. On some nights, you would not know it by watching how the team worked.

We were constantly chasing radio traffic, domestic disturbances, unauthorized vehicles, noise complaints, and reports of suspicious activity. The more we reacted, the more calls arrived. The team grew accustomed to bouncing from one urgent task to the next.

At first, I thought we were doing well. Response times were fast, and the briefing was well-received by the higher-ups; however, something was amiss. The logs were full. Everyone looked busy. But over time, I noticed something.

There was no proactive presence with the community. No visibility patrols were conducted. No follow-up investigations occurred. Junior soldiers were burning out, senior noncommissioned officers were merely triaging chaos, and relationships with tenant units were deteriorating. Relationships were not being built because the team only appeared when problems arose. During a staff meeting, the law enforcement director stated, "I do not believe the team understands its purpose."

That was a moment of realization. The team took a step back and rebuilt the shift plan from scratch. We analyzed which calls truly required a response and which ones could be delegated or prevented with better systems. We reintroduced visibility patrols, engaged unit leadership more directly, and made it clear: *We are not just a reaction force; we are a stabilizing presence on this installation.* This approach transformed operations.

The team did not merely respond to reports; it began preventing them. Morale improved because soldiers felt part of a strategic plan rather than functioning solely as a cleanup crew. The team and resources remained the same, but the impact was entirely different.

Busyness Is Not the Mission. Clarity Is.

What started as a high-tempo, high-response team became a crew of competent people chasing noise. We were answering the radio, not answering the mission.

We were not failing because we did not care; we were failing because we were not aligned. We were confusing activity with impact. And without realizing it, we had drifted. That is the danger of leading by urgency.

It is subtle at first. Be proud of the hustle. Admire the responsiveness. Over time, however, it becomes clear that effort is being expended in the wrong direction.

How Do You Get Back on Course?

- **Reconnect the team to the mission.**

  Begin with purpose. The team needs to know what success looks like beyond "get through the shift." Clarify why each member matters and how their actions align with the bigger picture.

- **Review the systems, not just the people.**

  Most performance issues are process issues in disguise. Is the team structured to be reactive or proactive? Rebuild the shift plan, patrol focus, and follow-up strategy.

- **Stop rewarding chaos.**

  Busy logs and rapid response times may feel like success, but do not confuse firefighting with fire prevention. Celebrate efforts that reduce calls; do not just respond to them.

- **Balance presence and performance.**

  High visibility fosters trust. Appear before problems arise. Cultivate relationships with stakeholders so they perceive the team not merely as a cleanup crew but as a leadership presence they can rely upon.

- **Protect your people from burnout.**
  Even the strongest teams will falter under continuous urgency. Schedule time for rest, review, and rhythm. Do not allow the team to operate on fumes. Empower them with structure, not just speed.

Mission Clarity Is Not a One-Time Speech; It Is a Daily Discipline.

When we stopped reacting and started leading with intention, everything changed. Same team. Same resources. But now, we were leading from alignment, not exhaustion.

Leadership is not about how fast you move; it is about knowing where you are going and bringing your people with you. Mission drift does not just break alignment; it breaks people. And when urgency rules unchecked, the next casualty is your team's energy.

Burned-Out Teams

In the early 2000s, **Nokia** was the undisputed leader in the mobile phone industry. They were on top of the world, enjoying a massive market share, global brand loyalty, and a history of innovation. But behind the scenes, Nokia's culture was deteriorating.

As smartphone technology advanced rapidly, Nokia's leadership became paralyzed by a toxic combination of short-term urgency and fear-based decision-making. The internal chaos appeared as follows:

- **Product launches** were rushed to the market to surpass competitors, often lacking strategic alignment.

- **Teams feared speaking up,** senior leaders had created a high-pressure environment in which bad news did not travel upward.

- **Every delay was treated like a crisis,** rustling in corners were cut and long-term development suffering.

- **Burnout was widespread,** engineers, developers, and project managers, juggling conflicting deadlines and priorities without clarity or consistent leadership.

Nokia *was not short on talent or innovation.* In fact, they *developed a prototype touchscreen smartphone before the iPhone.* However, the environment was reactive, filled with frequent fire drills and executive panic, preventing the device from reaching the market.

By the time Nokia attempted to adapt, it was too late.

In 2013, Nokia sold its mobile phone division to Microsoft.[1] A company once valued at **$250 billion** was reduced to a cautionary tale of leadership failure.

Your people do not burn out from working hard; they burn out from working in chaos, as the Nokia story illustrates. When every day feels like an emergency, each meeting is a fire drill, and every email is marked as urgent, your team depletes its emotional energy.

They cease taking initiative because they are in survival mode. They stop thinking ahead because there is no time for tomorrow. There is no feedback because "we do not have time for that right now." Eventually, they stop caring.

When that happens, performance declines, morale diminishes, and you become the leader of a team that is physically present but emotionally disengaged. Chaos is not a strategy, and urgency is not leadership.

Nokia did not collapse because it lacked innovation; it collapsed because it lost its alignment. They confused busyness for progress, urgency for importance, and fear for leadership.

If you are not careful, the same outcome can occur within your organization.

As a leader, it is your job to create clarity, not chaos, and to protect long-term vision from being sacrificed on the altar of short-term urgency. No matter how talented your people are, if the culture burns them out, you will not merely lose productivity.

You will lose your future.

How do you fix it if you are already there?

---

1. **"Nokia Sells Mobile Phone Business to Microsoft,"** *BBC News*, April 25, 2013, https://www.bbc.com/news/business-22286404.

Establish Psychological Safety

People need to know they can raise issues, share bad news, or challenge direction, without fear of backlash. You cannot lead what you cannot see, and your team will not speak the truth in a culture of fear.

Establish Priority Clarity Daily

Begin each day or week by identifying the top one or two priorities. Make it clear what matters *most* so that people are not whiplashed by every incoming fire drill.

Slow Down in Order to Speed Up

Do not rush to launch. Do not ship incomplete work merely to meet a deadline. Cultivate a culture that values thoughtful execution over reactive movement. Sound process surpasses hasty panic.

Protect Your Team's Time

Audit your meetings, eliminate unnecessary approvals, and enforce email boundaries. If your people are always reacting, they will cease creating.

Lead with Calm, Not with Crisis

Your team takes its emotional cues from you. When you lead with clarity, consistency, and confidence, even under pressure, they follow suit.

Urgency Is Unavoidable. Chaos, However, Is Optional

No Time for Growth

In the 1980s and early '90s, **IBM** was *the* technology powerhouse. The company dominated the computer market and was regarded as one of the most respected companies in the world. However, behind the scenes, cracks were forming, not because the company lacked the talent or resources, but because it was too consumed by short-term execution to recognize long-term disruption.

IBM employed engineers and developers who warned about the forthcoming wave of **personal computing**. The company had prototypes and internal advocates promoting innovation.

However, those voices were drowned out by urgency:

- Hitting quarterly numbers.

- Satisfying shareholders.

- Maintaining legacy systems and large client contracts.

- Competing with existing rivals in *their* space, not where the market was headed.

While they concentrated on keeping their mainframes operational and maintaining sale figures, **smaller, more agile companies** such as Microsoft and Apple moved swiftly to dominate the personal computing market.

By the time IBM tried to pivot, the market had already advanced. They did not fail because they lacked intelligence; they failed because they were too reactive. They were **busy**, not **strategic**.

Growth, whether personal, professional, or organizational, requires space and time. It requires margin to plan, mentor, train, and reflect. However, when there is a constant state of urgency, that margin disappears.

Strategic planning becomes a "someday" task, and, to my knowledge, "someday" cannot be found on any calendar. Team development is deferred. Process improvement is always postponed until after the next crisis.

Before long, your organization becomes stale. You are stuck in the same cycle of survival, and when the competition evolves or the environment shifts, you are the last to adapt.

Business Leadership Takeaway

Urgency may feel productive. However, growth resides in reflection, reinvention, and forward thinking. If you do not make time for strategic focus, another will, and the competition will outpace you to the future. **You do not just lose momentum; you become irrelevant.**

Apply the Eisenhower Matrix

1. Urgent & Important (Do Now)

These tasks require immediate attention and have a direct impact on the organization.

Examples

- Compliance issues

- High-level client problems

- Security breaches

Leadership Insight

If you are constantly in this quadrant, it is likely due to a lack of planning or ineffective delegation earlier on.

2. Important but Not Urgent (Decide)

This is where authentic leadership and growth happen. These tasks do not shout for attention; however, they shape the future.

Examples

- Strategic planning

- Mentoring team members

- Improving processes

Leadership Insight

Great leaders **schedule** time for this quadrant. It is where momentum is established and long-term value is created.

3. Urgent but Not Important (Delegate)

These tasks may appear urgent; however, they do not require your direct attention. They frequently interrupt focus.

Examples

- Routine emails

- Calendar logistics

- Low-risk issues

Leadership Insight

Train your team to handle these. If you do not, your calendar will get hijacked by others' priorities.

4. Not Urgent & Not Important (Delete)

These tasks provide little to no return and only consume time.

Examples

- Busywork

- Office gossip

- Endless reply-all email threads

Leadership Insight

Eliminating these tasks creates the margin necessary for deep work, reflection, and growth.

Why the Matrix Matters in Leadership

- It helps you lead with intention, not reaction.

- It brings clarity to chaos.

- It creates margin for innovation and strategic thinking.

- It prevents burnout by keeping your calendar aligned with what matters most.

The truth is that urgency will always pursue you, yet priority is a choice you must make. Outstanding leadership is not measured by how many fires you put out; it is measured by how many fires you prevent. It is measured by how effectively you prevent them, how clearly you stay focused on what matters, and how consistently you protect the space necessary for growth. Anyone can be busy; however, not everyone has the discipline to say, "These matters more." Leaders who change organizations, build trust, and leave a legacy are those are the ones who do not just respond to what is loud. They invest in what endures.

The Challenge

Before tomorrow starts, identify the top three things that truly move your mission forward, and protect them like someone's future depends on it. Because one day, it will.

Success Redefined: Personal Battles, Corporate Lessons, and the Discipline to Finish

Success is one of the most overused and least defined words in the leadership lexicon. Ask ten people what it means, and you will get ten different answers. For some, it is a personal journey, the pursuit of freedom, purpose, or stability.

For others, it is organizational growth, profit, or impact. But no matter how you define it, success does not happen by accident. In this chapter, we will explore what success means personally and corporately, then walk through two tools every leader needs to pursue it: planning backwards and taking relentless action.

Define Success

Defining success is the starting line for a long journey in leadership, but it is personal experiences that shape how we truly understand it. What does success mean to you? You cannot achieve what you cannot truly define.

For some, success is wealth. For others, it is peace, purpose, or freedom. The true answer to success will depend on whom you ask. If you look up the word "success," most definitions will describe it as the accomplishment of an aim or purpose.

To achieve success, the first step is essential yet straightforward: do you know your aim? Success often has several common threads. First, it begins with your definition: What does success look like for *you*? Second, you must understand that it evolves. As you grow, or as your organization evolves, your vision of success will also change. When you ask whether success is an endpoint or an ongoing pursuit, the truth is that it will remain a continuous battle.

Defining success is the starting line, yet personal experiences shape our proper understanding of it. Let me give you an example from my journey that made me reconsider everything I thought I wanted.

Personal Success

When I transitioned out of the military, I expected the corporate world to welcome me with open arms because of my background. I had the experience, the titles, and the leadership credentials. I believed that the right position with a six-figure salary would appear quickly. It did not.

I applied everywhere. I maintained an Excel spreadsheet that tracked all the positions for which I applied and how many responded with the standard rejection letter stating, "Even though your résumé was excellent, we chose to go in a different direction." Weeks passed. I checked the calendar, and my leave days had expired. My final Army paycheck was approaching, and I still had not received any offers. I had assumed that securing a six-figure salary after leaving the Army would not be as challenging. I was mistaken.

One day, I received a telephone call. It was for a leadership position with a highly reputable company. The salary was excellent. The title was impressive. The only remaining step was to travel to the headquarters for final interviews. It was the type of position that would cause individuals on LinkedIn to double-click with envy. However, as I approached the finish line, something felt amiss. I arrived early, sat in the parking lot, and watched people entering the building. There was no eye contact. No smiles. No small talk. Then, during check-in, I bumped into a guy in the hallway. I apologized; he was cordial and we had a quick conversation about what I was doing in the building, wished me luck, and suggested that I stop by before leaving. The interaction happened so quickly that I did not catch his name. When I asked the human resources representative for his identity, they had no idea.

That stuck with me. How can you not know a colleague who works right in the same building and on the same floor? The office looked good. The role seemed polished. However, the culture felt hollow. I asked myself the question that too many people avoid: "Am I excited about the actual work or just the title?" I received the formal offer the next day upon returning home, and I declined it.

Later, I accepted a different role with better people and stronger values, but still, something was absent: the freedom to build, coach, and lead from the front. It did not align with my personal approach. That experience guided me to my current work, which involves assisting individuals in leading rather than merely occupying leadership positions. That was my personal definition of success. It evolved as I evolved. Just as individuals struggle to define success, organizations must do the same. But the stakes are higher because their answer shapes entire cultures, strategies, and futures.

Let us look at how companies define success and how those definitions drive everything they do.

Corporate Success

For some companies, success signifies growth and profit. For others, it is about purpose and making a meaningful impact. Consider two highly contrasting examples: Amazon and Patagonia.

When Profit Is the Point: Amazon

Amazon defines success in terms of market share, speed, and operational superiority. From the beginning, Jeff Bezos emphasized customer focus and rapid reinvestment.

Everything at Amazon is measured: delivery times, pricing efficiency, and ecosystem growth. It is a success model built on scale, speed, and relentless execution. It works, but not everyone agrees with the associated cost.

When Purpose Leads: Patagonia

Patagonia defines success in terms of environmental and social accountability. Its founder, Yvon Chouinard, famously ran a 2011 Black Friday ad to ask customers not to buy a jacket.

In 2022, he transferred ownership of Patagonia to a trust and nonprofit to combat climate change, stating: "Earth is now our only shareholder."

Patagonia measures success in terms of impact, not just income. It invests in fair trade, minimizes environmental harm, and supports activism. Growth is not the goal; legacy is.

Whether your vision of success looks like Amazon's relentless drive or Patagonia's purposeful stand, neither outcome occurs by chance. Behind every definition of success exists a process, a way to translate vision into reality. One of the most potent tools I carried from the military into business was backward planning.

Whether you chase profit like Amazon or impact like Patagonia, reaching your version of success does not occur by accident. It takes intention, and one of the most powerful tools I have learned, both in the military and business, is backward planning.

Plan Backwards

Most people plan forward: "What do I do first?" or "What is next?" But *great* leaders start at the final objective. They define the destination and then work backward to plan the steps to get there. Consider the process of catching a flight.

The flight leaves at 7:00 a.m. You need to be at the airport by 5:30 a.m., so you need to leave your house by 4:30 a.m. Therefore, you set your alarm for 4:00 a.m. This is backward planning.

In the military, we used this approach constantly. We asked, "What does success look like?" and worked backward from that end state. It eliminated guesswork and established alignment. In business, the same logic applies. Want to launch a new product by October 1st? Then work backward:

- Marketing must go live by mid-September

- Manufacturing must be completed by August

- Packaging must be finalized by July

- Retail contracts must be secured by June

- Team kickoff must occur in May

The outcome dictates every step.

Backward planning transforms your goal into a roadmap. Everyone understands the objective. Everyone understands their role. There is no more guessing. There is no more scrambling. And here is the truth: you have done this your whole life. You simply have not named it.

Having a plan provides direction, but a plan without action is just words on paper. What separates dreamers from doers is the execution of showing up consistently, even when it is hard. That is where relentless action becomes essential.

Having a plan organizes you. However, having a plan does not guarantee progress. What separates dreamers from doers is the discipline to execute that plan consistently, especially when it is not easy. That is where relentless action becomes essential. At every military leadership school I attended, the instructors would place us in stressful situations, forcing us not only to devise a plan but also to make decisions and take relentless action that was aligned with our personal or organizational goals.

Implement Relentless Action

This is where most people fail. Not in the idea itself. Not in the strategy itself. Instead, it is in the execution. They brainstorm, meet, use a whiteboard, and utilize a vision board. Then they wait idly. For the perfect timing to occur. For complete clarity. For sufficient confidence.

Meanwhile, the plan accumulates dust. But success is not about having the perfect plan. It is about movement. Not just any movement, but relentless, aligned action. This is important not only in your professional work but also in your personal life.

Relentless action does not mean burnout. It means consistency and persistence. It means showing up and completing the work, even when you do not feel like it. Most of us understand that our minds are wired to do the things that provide immediate self-gratification, and relentless action fits this mode sometimes, but other times it does not. Even when motivation is absent, it means not letting a bad day become a bad week.

Aligned action means every step ties back to the goal. In your personal or professional life, you encounter two types of people: those who help you reach a goal and those who distract you from it. Busyness does not equal progress. Some people who work late often share the news with others as a mark of honor. This does not constitute a badge of honor; it is not aligned with your goals. You can be full of motion and still advance nowhere.

When I was in company command at Guantanamo Bay, I led a Headquarters Company that provided support for detention operations. We were not directly securing detainees, but we were the driving force behind the mission.

Every military unit has a METL (Mission Essential Task List). These lists represent the absolutely necessary tasks. These are the non-negotiables. If you cannot accomplish them, the mission fails. In business terms, they constitute your core capabilities.

I trained support staff on tasks they thought they would never use. It was not easy. Motivation declined. I stayed aligned and consistently pursued action during our training events. Then, one day, the unexpected happened. Those same staff were required to put on riot gear and restore order. On that day, all the training proved its value. That is relentless action. It is not glamorous or exciting. Still, it is necessary.

Consider a rocket. The launch requires a massive push, but after that, it is about staying on course. That is how success happens. You will not always feel inspired. That is everyday life. Still, discipline is more reliable than motivation.

You have defined success. You have planned backward. Now comes the part that separates those who say from those who do: you must take action. Relentless, aligned action is essential. This is the distinguishing factor.

Summary

Define What Success Means to You and Your Organization

Takeaway: Leaders must define what success means *before* they can effectively pursue it. **Without a clear aim, even the most diligent leader may fail to achieve it.** Your people need a vision they can see and believe in.

How to Achieve It

- Reflect personally: Is your success defined by money, purpose, freedom, legacy, or something else? Not only should you know, but your team should also.

- Revisit and refine: Allow the definition to evolve as you grow or your company matures.

Success Is Not Static—It Evolves

Takeaway: What success meant to you five years ago should not be the same today. If it is, you are not growing. Leaders often chase yesterday's goals out of habit instead of reassessing what drives them now.

How to Achieve It

- Set time regularly to review your why.

- Seek feedback on your leadership impact from trusted peers and colleagues.

- Reevaluate your definition of success based on impact, not personal pride.

Plan Backwards to Move Forward

Takeaway: Do not guess your way to success. Reverse-engineer it. Backward planning creates clarity, accountability, and momentum. It ensures that daily actions serve long-term goals.

How to Achieve It

- Begin with a clear end goal that is specific and measurable.

- Identify milestones that must be achieved along the way.

- Assign timelines and allocate resources starting from the deadline and working backward to the present day.

Take Relentless, Aligned Action

Takeaway: Discipline surpasses motivation in every circumstance. Relentless action, especially when no one is watching, is what moves leaders and teams forward. Most leadership failures happen in execution, not strategy. Ideas without action are merely thoughts.

How to Achieve It

- **Establish habits that support your goals,** including daily check-ins and non-negotiable routines.

- **Measure output, not mere busyness.** Your team should report that they were productive, not busy.

- **Ask daily:** *Is what I am doing aligned with the desired outcome?* If not, stop doing it.

Train for the Unexpected

Takeaway: Do not just train for what is likely; train for what is critical. It can be something as simple as a fire drill, a power outage, or an active shooter drill. No one likes doing these drills because they interrupt workflow, but if you ever have to do it, you will be glad you trained on it.

How to Achieve It

- Identify your team's METL.

- Rehearse worst-case scenarios, not just ideal outcomes.

- Reinforce that preparedness is a form of leadership love; you are investing in their survival and success.

Conclusion

The Choice Is Always Yours

Failure is never the end of the story. It never has been. Failure is a moment, a data point, not a definition. And yet, for too many people, it becomes the excuse they hide behind, the justification they cling to, or the reason they stop reaching for more. They convince themselves that falling short means they weren't meant for leadership, weren't cut out for responsibility, or weren't capable of more. That belief is wrong. What defines us personally and professionally is not the moment we fail. It's what we do *after* we fail. It's about being honest, taking ownership, and being willing to do the hard work required to move forward. Everyone fails. Not everyone fights. And far too few will finish.

Throughout my career, I've watched talented, capable leaders walk away from their potential not because they lacked ability, intelligence, or opportunity, but because they quietly quit on themselves. They stopped holding the standard. They stopped addressing matters early. They chose comfort over growth and familiarity over excellence. They didn't fall; they settled.

You don't have to make that choice.

Failing is inevitable. Fighting is intentional. Finishing is a decision you make again and again, long after motivation fades and when no one is watching. This book was never about avoiding failure. It was written to prepare you for it. Along the way, I've shared lessons, experiences, and practical suggestions that are not theory; they are things I had to learn to work through, and I wanted to assist you on your path as a leader, not conceptual ideas or trendy slogans, but advice grounded in real leadership, real consequences, and real responsibility.

These leadership lessons apply no matter where you are right now. Whether you are leading yourself, entering your first leadership role, responsible for a small team, or accountable for an entire organization, the fundamentals persist the same. Leadership is not defined by title, rank, or authority; it is defined by behavior, consistency, and character.

If you are waiting for a title before you act like a leader, you will always be waiting.

Leadership begins the moment you decide to take ownership of your standards, your actions, and your outcomes. But allow me to be clear: reading this book alone will not change anything.

Awareness without action leads right back to the same results. Insight without discipline turns into good intentions and lost chances. Growth only happens when you apply what you know, especially when it's uncomfortable. So this is where it gets real.

This is where you draw the line and decide what you will and will not accept from yourself and from others. This is where you commit to dealing with problems early, rather than hoping they'll fix themselves. This is where you choose consistency instead of convenience, courage over comfort, and discipline over excuses. You will fail again. That's not pessimism, it's reality.

The difference going forward is this: when failure shows up, you will recognize it for what it is, a test. A moment that demands a response. You can avoid it. You can excuse it. Or you can fight through it and finish what you started. The leaders who succeed aren't the ones who never fail.

They are the ones who refuse to stay down. The ones who learn, adapt, and take responsibility. The ones who understand that leadership is not a final point, but a daily commitment.

So take the lessons in this book. Apply the suggestions. Hold the standard. Lead where you are, with what you have, right now. Because the next chapter isn't in these pages. It's written in your decisions. And when it matters, when pressure increases, when expectations rise, and when quitting would be easier, make the choice that defines real leaders.

Fail. Fight. Finish.

The choice is yours.

Glossary of
Abbreviations

Glossary of Abbreviations

AARs: After-Action Reviews

ASVAB: Armed Services Vocational Aptitude Battery

BC: Battalion Commander

BCT: Brigade Combat Team

BSTB: Brigade Special Troops Battalion

CEO: Chief Executive Officer

CG: Commanding General

CSM: Command Sergeant Major

CTA: Chicago Transit Authority

DOOO: Differences of Organizing Opinions

EL: Elevated Train

EOC: Emergency Operations Center

FOB: Forward Operating Base Shank

GT: General Technical score

HR: Human Resources

MASCAL: Mass Casualty Training

MDMP: the Military Decision-Making Process

METL: Mission Essential Task List

MOS: Military occupational specialty

MP: Military police

NCO: Noncommissioned Officer

OCS: Officer Candidate School

PCS: Permanent change-of-station orders

PL: Platoon Leader

POTUS: President of the United States

PTO: Paid time off

PT: Physical training

QRF: Quick Reaction Force

RASP: Ranger Assessment and Selection Program

ROE: Rules of Engagement

SAT: Scholastic Aptitude Test

SOP: Standard Operating Procedure

SSG: Staff Sergeant

XO: Battalion Executive Officer